The Fox, the Shrew, and You

The Fox, the Shrew, and You

HOW BRAINS EVOLVED

ROGIER B. MARS

PRINCETON UNIVERSITY PRESS

PRINCETON AND OXFORD

Published by Princeton University Press
41 William Street, Princeton, New Jersey 08540
99 Banbury Road, Oxford OX2 6JX

press.princeton.edu

GPSR Authorized Representative: Easy Access System Europe - Mustamäe tee 50, 10621 Tallinn, Estonia, gpsr.requests@easproject.com

All Rights Reserved

ISBN: 9780691238920
ISBN (e-book): 9780691238937

Library of Congress Control Number: 2025945785

British Library Cataloging-in-Publication Data is available

Editorial: Alison Kalett and Laura Lassen
Production Editorial: Jaden Young

Jacket/Cover Design: Chris Ferrante
Production: Danielle Amatucci

Publicity: Matthew Taylor and Kate Farquhar-Thomson
Copy Editor: Lucinda Treadwell
Jacket images: Adobe Stock

This book has been composed in Arno Pro.

Printed in the United States of America

10 9 8 7 6 5 4 3 2 1

CONTENTS

The Fox, the Shrew, and You

Introduction

DIFFERENT BRAINS

Different brains are everywhere around us. And they fascinate us.

Which parent of a young child has not wondered what happens inside that little head? And been delighted when the child suddenly produced a new behavior—a fascination with numbers, understanding how to twist jigsaw pieces to fit, or simply standing up—like a light went on?

Which dog owner has not pondered how much the animal truly understands? How similar the dog's thought processes are to our own?

We wonder about the brains of other people constantly. Their thought processes, whether they perceive the world exactly the way we do, what they know, think, and feel.

And then there are the truly alien brains. The tiny ants that are able to produce complicated colonies, some even using other animals as slaves. The octopus with its distributed processing centers in its eight limbs.

How do all these brains relate to our own?

This book aims to answer these questions. As any biological system, the brain is a product of evolution. Through descent with

1

modification from a common ancestor each brain in the world today both shares some aspects with all others and is truly unique. By comparing brains and understanding what different circumstances occurred during their evolution, we can start to understand both our own brain and its relationship to other brains.

Science is currently going through a revolution in our understanding of the brain. In these pages I hope to share some of the new insights we have recently gained and—probably—shatter a few myths along the way. But before we do this, let's make sure we are all familiar with the main ingredients: evolution, comparative biology, and modern neuroscience.

―――――

Thinking about brain evolution

January 1871. Charles Darwin is sitting in his study in Down House, the manor in the Kent countryside where his family moved to get away from busy London life. He has just finished correcting the proofs of his new book, *The Descent of Man*. In this work, he explores whether the rules for evolution by natural selection apply to humans as much as to any other living being. It is a book he never intended to write. Although Darwin was a prolific writer, he was very cautious of revealing too much of his theories before he felt they were absolutely ready. When he first published his theory on natural selection, it was under pressure. A letter he received in 1858 tipped the balance. The explorer Alfred Russell Wallace had reached similar conclusions to Darwin and asked Darwin to forward his letter outlining his theory for publication. The fact that somebody was about to scoop him pushed Darwin to publish his ideas on natural selection. He always suspected his theory would be explosive. To keep it a bit under control, he refrained from mentioning what his theory meant for humans—that they were not the pinnacle of God's creation, but simply another life form among many.

The central problem that Darwin tried to address in his work was whether species are stable or can change over time. The Bible of course states that all species have been created in their present form, but the idea of gradual change—evolution—was already present in Victorian scientific thinking. During his voyage around the world on the HMS *Beagle*, Darwin noticed fossils of extinct animals that very much looked like the present-day animals living in similar locations. The present-day animals seemed to have evolved from the fossil ones. The problem was how did they change and why? Some people had proposed that changes occurring during life are passed on to the next generation; some favored a more directed evolution with solutions to problems appearing spontaneously in some individuals. Darwin agreed that variation was key. But he did not agree that variation could be acquired and passed on, or that variation had a goal in mind. The crucial element he added was selection. If there is random variation in the population, and some variations make you more likely to survive, then that means that some of the variations are simply more likely to spread. Darwin's crucial insight was thus not that evolution happens, but that adaptation occurs by natural selection among existing variants. This explains why a certain trait is present. At some point along the evolutionary path, a trait helped its owner survive or, more accurately, have more surviving offspring. No creation by God, no selection of acquired characteristics, no prospective insight of evolution. This was the theory he published in his seminal work, *On the Origin of Species* in 1859. Anticipating that his theory would stir up trouble, he decided to hardly mention humanity in his book.

Since the publication of his theory in 1859, things had changed. Sure, there had been uproar and fierce discussion, most famously at a debate at Oxford's Natural History Museum where the bishop Wilberforce is believed to have asked the Darwinists if they descended from apes on their mother's or father's side. But there had also been acceptance. In the beginning of his new book, Darwin explained how young researchers were increasingly accepting of his theory. The time was now right for him to systematically investigate how applicable his ideas were to humans.

It is worth thinking a bit about what Darwin's idea means. It means that to understand any aspect of life—be it an organism, an organ, or a behavior—we need to understand its ancestral history and the drivers and constraints of change. To start with the ancestral history, this implies that related animals are likely to be similar—their bodies and organs did not appear spontaneously from nowhere. This is why animals have rudimentary organs, like the muscles some of us have to move our ears. They are leftovers from something that our ancestors had. It means there are constraints. Selection can only work on the natural variation that occurs in the population, so changes are likely to be gradual. We also have to work within the context of an already functioning organism. We do not just grow an extra arm, because that would not be compatible with the existing body plan. Most importantly, it means that for a trait to have made it into the population, it should have given our ancestors some benefit, some advantage over those who did not have it. The advantages themselves are shaped by the environment of that ancestor. They have to work in those circumstances, without foresight. Importantly, adaptations at one moment in time can seriously constrain an animal in the future. An extreme example of that is the extinction of the dodo. Being a trusting non-flying big bird can be fine if you live on an isolated island, but works a lot less well when large human predators arrive—the dodo ended up being an easy dinner for hungry sailors. Natural selection for trusting dopiness worked well at some point in time, but less well at another point. So, to really understand why something is the way it is in biology, we need to take into account its adaptive benefits at the time that it appeared, not just its current function.

In the early versions of *The Origin of Species*, Darwin did not really talk about brains. That changed in this new book. In building his argument for continuity between humans and other animals, he referenced work describing how similar the human brain looks to that of the orangutan. He then described many aspects of the brain's output—behavior and mental abilities—that he believed are similar between humans and nonhuman animals. He conceded that humans are particularly good at some things, like using tools, but argued these were mostly differences in degree rather than in kind. He also argued that such human abilities vary in the population, meaning there is variation that can be the target of selection. The brain, he argued, is subject to the rules of evolution by natural selection, just like any organ. Charting brain evolution was not Darwin's goal, however. He gave us all the ingredients to find out about brain evolution, but we still have to bake the cake.

Comparing brains

Autumn 1891, the island of Java in the then Dutch Indies. Eugene Dubois is directing his team of mostly convicts digging in the ground of a riverbank. Dubois had left a cushy position as lecturer at the University of Amsterdam, took on a position as army physician, and dragged his wife and newborn daughter out to the colonies, all to pursue his dream—to find the missing link between humans and great apes. He agreed with Charles Darwin that humans evolved from a common ancestor we share with the living apes—chimpanzees, bonobos, gorillas, and orangutans. He disagreed with Darwin on where this ancestor had lived. Where Darwin suggested Africa, Dubois bet on the tropics of Asia. His efforts would be rewarded. His men found a tooth and a skull cap, the top bone of the head. It had a heavy bow ridge. It was not an ape, but it was not quite human either, while the tooth looked more like that of an ape. A year later, Dubois's men found a femur, also known as a thighbone, about 15 meters from where the skull was found. Its shape suggested its owner had walked upright, just as modern humans do. From the skullcap of the head, it was possible to estimate the size of the brain that was housed in this head. It turns out it was about 900 grams, smaller than the 1.3 kilograms of modern humans, but much bigger than the 400 grams of the chimpanzee. Dubois was convinced he had found his missing link and termed the creature *Pithecanthropus erectus*, the upright ape-man. Today, the species is called *Homo erectus*. It lived between two million and 100,000 years ago and was the longest-living species of human ever to walk the earth.

If we want to know something about the bodies of our ancestors, one way is to go out and look for fossils, as Dubois did. Fossils are any preserved remains, impression, or trace of a once-living organism from a past geological age. They can be stone imprints of microbes, mosquitos preserved in amber, or DNA remnants, among others. But in most cases, what people associate with fossils are the petrified bones of animals. Unfortunately, soft tissue such as brains generally does not fossilize.

This leaves paleontologists with only very indirect measures of brains gone by. If a skull is found at a fossil site, it is possible to use it as a mold to create an endocast to get a rough estimate of the size and shape of the brain. We can then compare the size of the brain to the size of the body. Larger animals tend to have bigger brains, but if an animal has a particularly large brain for its body size, we say it is encephalized. This might mean that the animal has invested in brain size as an evolutionary adaptation. The study of the relationship between brain size and body size is part of the scientific field of allometry. Perhaps unsurprisingly, Dubois was one of the pioneers of this particular field. Upon his return from the tropics, he took up a position at the Teylers Museum for art, natural history, and science in the city of Haarlem in the Netherlands. At the time, the museum housed one of the most spectacular collections of animal skeletons in the world. The variety of fossil skulls allowed him to study encephalization of the brains of different mammals.

But the approach of using the skull to reconstruct the size of a brain assumes the entire space in the skull was previously filled by the brain, which is not always true. In many early mammals, the brain was kept in place by cartilage, another tissue that does not fossilize. Endocasts also only tell us about the outside shape of the brain, nothing about its cells and its internal organization. Some big brains contain folds, only some of which might leave an imprint on the skull that would in turn be visible in the endocast. If we know a bit about how different parts of the brain relate to such landmarks, it can help us reconstruct some of the brain, but it will always be an approximation. In all, fossils will always provide us with only limited information about their owners' brains.

If fossils are so unreliable, what else can we use? The solution is to compare the brains of living species. Contrary to the

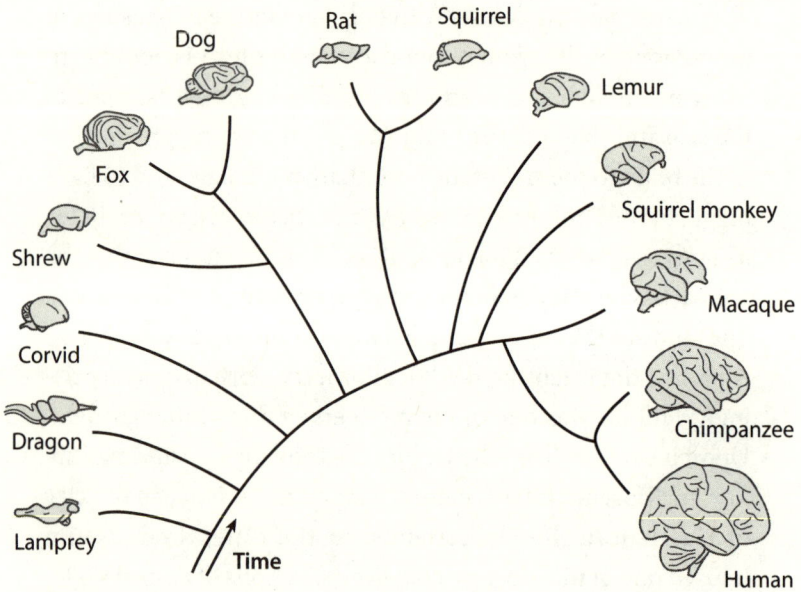

FIGURE 0.1. Evolutionary tree showing the vertebrate brains discussed in this book.

famous picture of human evolution where a chimpanzee slowly morphs into a modern human over evolutionary time, all currently living species are what we call crown species. This means that all animals are the current representative of a unique lineage of them and their ancestors back to their most recent ancestor with another living species. In other words, all currently living species are to some extent equally evolved and none are the ancestors of others. We have a common ancestor with chimpanzees and this ancestor was likely a great ape; chimpanzees are not our ancestors but our sister species. By systematically comparing similarities and differences between living species, it is possible to reconstruct their ancestral relationships. Traditionally this was done by looking at similarities and differences in the size and shape of different organs or similarities and differences in developmental patterns. Nowadays this is mostly done through comparison of genetic material. Animals whose genetic material is more similar are assumed to have diverged from a common ancestor more recently than animals whose genetic material differs more. Using this approach, we now have a pretty good idea of the evolutionary tree of living animals. We can use this understanding of the evolutionary tree to investigate how the brain has changed. If we compare the brains of, say, a human and a chimpanzee, any feature that they share is thought to have been present in their common ancestor. If we find anything that is different, something changed in the lineage leading up to the human or in the lineage leading up to the chimpanzee. To find out which it is, we can look at an additional species. Gorillas had a common ancestor with humans and chimpanzees about nine million years ago. If we find something that is different between humans and chimpanzees, but is shared between gorillas and chimpanzees, we can infer it is due to a change in the line leading up to humans. If we then would also know how the environment of early humans differed from that of chimpanzees and gorillas, we can start figuring out what the role of the human feature was.

What about when very distinct animals share a particular feature? For instance, humans, elephants, and dolphins all have very big brains for their body sizes, but they are not very closely related. In fact, all of them have much more closely related animals with much smaller brains. This could be a case of convergent evolution, where a similar feature has evolved multiple times in unrelated species. Convergent evolution is really useful, because we can use it to link the brain to its environment. If the feature not only evolved multiple times, but it did so under similar circumstances, we have a much better idea what problem that part of the brain helped to solve.

Ultimately, we want to understand what changed in the relationship between an animal and its environment, its *ecology*. If an animal's environment changes and it needs new strategies to survive, variations that occur in the population might mean that some individuals start being more successful. Differences between individuals in bodies and brains—they often go together—will mean that selection can do its work. We should keep in mind, though, that we are looking at animals that are the results of long evolutionary histories, even since the time of their most recent shared ancestor with another species. Humans did not suddenly appear from a common ancestor with chimpanzees due to a single change in ecology. More likely, two populations of human-chimpanzee ancestors got separated, the circumstances for the two populations changed in different ways, and step by slow step, different adaptations got favored in the two populations. The species that survive today, we call "chimpanzee" and "human," but they are the surviving end results of many experiments to adapt to circumstances over time.

The comparative method gives us a way of studying brain evolution when fossils are not available. If we can analyze enough brains, find what's similar and what's different, and relate that to the animals' ancestral history and their current circumstances, we can get an idea of what was driving their evolution.

———

Scanning brains

Summer 2016, a basement in Oxford. My colleague Sasha and I are sitting behind a series of computer screens, staring at images coming in. From behind a shielded door comes the sound of the MRI scanner. MRI scanners are standard in every hospital, used to look at the inside of people's bodies without the need to open them up or to expose them to potentially harmful X-rays. The technology has revolutionized medical assessments. The MRI scanner in our basement is smaller than the standard hospital issue. It is optimized to scan rodents. Mice, rats, and other small rodents are frequently used in medical research to test ideas before they are applied to humans. What is in the scanner today, though, is not a mouse or a rat. It is the brain of a macaque monkey. Sasha and I concentrate on the screens as the first image of the brain becomes visible. "Looks good," he says.

Comparing brains has always been a laborious endeavor. Brains are not a uniform whole. Different parts of the brain have different types of cells and group cells together in different ways, forming distinct regions. To really understand a brain, we need to map out the different brain regions. Before MRI, this meant lots of painstaking work. To map out brain regions, one had to get a brain and slice it into very thin slices. The slices were then treated with various chemical compounds that make different structures of the cells visible. Then one looked through a microscope to see how the cells are organized. When the pattern of organization changed, this would indicate that we had moved into a new brain area. This was done by hand, slice by slice, all through the brain. For a brain as large as the human, this could take years.

Taking years to study a single brain is not feasible in today's scientific funding landscape. Thankfully, biology is living through a time of technical revolutions, giving rise to a new generation of comparative scientists. In neuroscience, techniques like MRI allow us to scan whole brains in minutes. The information they provide is not quite at the level of the cells under the microscope, but it is getting closer all the time. The technical revolution is nowhere greater than in genetics. Whereas the deciphering of the first human genome took 13 years and $300 million, while I write it is now closer to 1 day and $1000. Probably when you read this, it will be less still. There are also high throughput methods that can determine where in the brain certain genes are expressed, allowing us to investigate how genes get switched on and off during development, including in the brain. In short, it is now possible to get lots of different types of information about brains in a relatively short period. It is the reason why I started contacting zoos to ask about their animal brains. Many neuroscientists are finding brains to study using the new approaches. Some acquire governmental permits to capture some animals, scan their brains, and release them again. Others focus on cadavers that are found in nature and studied by governmental agencies for pathogens. Others, including me, work with zoos. Many zoos have a research mission. They want to learn about the magnificent animals they care for. So, if after the animal dies we can learn even more about them by saving their brains, spines, or eyes, they contribute. Without them, a lot of my research would not have been possible. It is how I ended up on that summer day scanning a monkey brain in an Oxford basement.

Plan of the book

The work of the new comparative neuroscientists, and the generations of researchers whose shoulders they stand on, means that we learn more and more about the wonderful diversity of brains we can find in nature every day. It means we can finally study where our brain comes from. In this book, we will take up that challenge.

We will use the comparative method to travel through the animal kingdom. In each chapter, we will compare two or three animals that are representative of larger groups—reptiles and mammals, rodents and primates, foxes and dogs, chimpanzees and humans, rodents and birds—and see what circumstances they lived in and how their brains adapted accordingly. We cannot trace the entirety of brain evolution in a single book. Therefore, we will concentrate on comparisons that are ultimately relevant to understanding our own human brain. This is not because we humans are in any way the pinnacle of evolution or that we are more evolved than any of the other brains we discuss. It simply because, as humans, we tend to mostly be interested in ourselves.

Because the brain is involved in so many complex behaviors, it helps to focus a bit on a basic function. I will argue in the first chapter that a useful way to study the brain is to view it as a foraging device, an organ that produces the behaviors needed to help us find the nutrients we need to survive. Different animals, we will see, need to solve different problems in their foraging. Studying these problems helps us understand differences in the organization of their brains.

In the first chapter, we will look at some relatively simple brains: the sea squirt, the lamprey, and the bee. We will use these simple brains to see where brains come from and what they are for. The lamprey has a brain similar to that of an early vertebrate, and we will use it to introduce some of the major subdivisions of the brain. In the second chapter we will look at two major groups of animals, reptiles and mammals, represented by the Komodo dragon and the shrew. They show us how animal life adapted to life on the land and, in the case of mammals, to produce a very flexible part of the brain. In the third chapter we will look at the group of animals that we belong to: primates. Comparing them with the most populous group of mammals, the rodents, we will see how the primate brain specialized for a lifestyle dominated by vision and fine movement skills. Within primates, some groups adapted to life in an uncertain environment, as we shall see in the fourth chapter. One type of behavior that has evolved a number of times in different groups of animals is sociality, the ability to live together with conspecifics. We still take this topic up in the fifth chapter and then see how this sociality influenced human evolution in the sixth chapter. Finally, in the last chapter, we will look at some other interesting brain adaptations and what they tell us about general patterns of brain evolution.

Throughout this journey, we will encounter strange brains, big changes in the world's climate, happy accidents, and wonderful behaviors. Let's go.

1

The sea squirt, the lamprey, and the bee

WHAT BRAINS ARE FOR

Lamprey

The sea squirt is not an attractive animal. It goes by the charming Latin name of *Ciona intestinalis* (pillar of intestines) and, quite frankly, that is not a bad choice given the looks of an adult sea squirt. The sea squirt lives at the bottom of the sea. It spends its adult life attached to a single patch. It is a filter feeder, meaning that it passes water through its body and filters out useful particles. Water is taken in and expelled through tubular siphons. In between, the animal's sack-like body filters nutrient particles out of the water. As a lifestyle, it is not very glorious, but it has kept sea squirts going for a good 518 million years.

The story of the sea squirt's life cycle is often used to describe why animals have brains. The sea squirt doesn't start life as a pillar of intestines. Its larva resembles a small tadpole, moving around the ocean floor by moving its tail. Once the larva is fully grown, it selects a nice spot on the sea floor and attaches itself there. The first thing it does in its new home is to consume its own brain! The brain, apparently, is no longer needed when the animal no longer moves around. The neuroscientist Daniel Wolpert jokingly likens this to an academic obtaining tenure: when you find a nice and cushy space to spend the rest of your days, you do not need a brain anymore. The point of the story is that the sea squirt's brain, primitive though it is, is essential for movement. If you don't move, you don't need a brain. That makes some sense. After all, plants don't move and plants don't have brains.

The story of the sea squirt's life cycle raises an interesting question: why have a brain? Think about it. As humans, our young take the better part of two decades to develop, mostly because their brains are still maturing. A young gazelle, in contrast, is up and running with its parents 30 minutes after birth. Helpless as human babies are, at the time of birth their brain is already so big that giving birth is generally the most dangerous thing a woman of our species can do. Before the rise of modern medicine, women frequently died giving birth, something that is a rare occurrence for our smaller-brained chimpanzee cousins. At the other side of life, our brain is vulnerable to strokes and an array of neurodegenerative diseases. During life, the brain is expensive to maintain, gobbling up nine times as much energy as the average organ. In fact, our ancestors might have had to invent cooking to outsource food processing, just to be able to extract enough energy to feed our hungry brain. Why would we have such a slow to develop, energetically expensive, and fragile organ?

I will argue in this book that our brains evolved to help solve a particular challenge: that of obtaining nutrients that are distributed in space and time. At some point in evolution, life forms on one branch of the tree of life invested in a strategy of obtaining energy-rich nutrients that were hard to obtain but promised great reward. They invested in brains to help them overcome the obstacles of obtaining these energy-rich nutrients. Of course, this "investment" was not a conscious choice. The law of natural selection means that some variations in life forms proved successful enough to survive and reproduce. Some variants for which this happened to be the case were those able to obtain difficult to get, but energy-rich, nutrients distributed across space and time. Another word for that behavior is *foraging*. I propose that the brain is a foraging device. And I will argue that differences between different animal species' brains largely reflect solutions to their foraging problems.

Does my theory hold in the case of the sea squirt that supposedly consumes its own brain when it settles down? It does. It is true that the sea squirt undergoes a substantial metamorphosis when it reaches the adult stage. It changes from a tadpole-like being into a creature with a sack-like body with two tubular siphons. During this metamorphosis, most of its brain cells do degenerate, although it is worth pointing out that there probably were only 177 neurons to begin with. But the sea squirt does maintain a kind of brain during its adult life, albeit one that is radically different from the one it possessed as a larva. One of the functions of this brain is to control the rate at which water is taken up and expelled, which depends on the food content of the water. If there are not enough nutrients in the water, the sea squirt will take in more water to compensate for the deficit. You could argue that, although the sea squirt itself doesn't change location, it does affect movement of the water. Just as it did when it moved itself, it has a type of brain that controls the movement of muscles. Those muscles move with a very specific goal, to ensure that enough water flows through its body to filter out the right amount of nutritious particles. In both larva and adult stages, the brain controls movement for a specific purpose: to ensure that enough nutrients find their way into its body.

But we need more evidence. Animals are the only organisms we know of with brains. If our proposal is true, then somewhere in the evolution of animals something changed in their foraging strategies. That change coincided with appearance of brains. Moreover, those brains must have provided an advantage to the foraging animal. In this chapter, we will examine each of these three arguments. First, we will chart the history of life on earth, up until the first complicated animals. Then, we will have a look at when brains cells and brains appear. Finally, we will see if these brain cells were in some way special, helping their owners forage.

A (very) brief history of life

The more than 500-million-year-old sea squirt is quite a recent player on the evolutionary stage. Life started early. The first cells, the smallest self-sustaining units of life, probably emerged about 3.8 billion years ago, when the Earth itself was a mere 750 million years old. Hydrothermal vents on the ocean floor provide the minerals and energy that made it possible for organic molecules to form. A crucial step in the evolution of life was the formation of a molecule that was capable of catalyzing other chemical reactions, including those creating more of itself. This self-replicating molecule was RNA, the initial genetic system; however, RNA on its own is vulnerable and likely to be destroyed. Life became truly viable only when RNA became enclosed by a protective membrane of lipid molecules. These molecules have one side that is soluble in water and one side that is not. Placed in water, these molecules spontaneously form two layers, because the insoluble parts line up toward one another. This way, a membrane is formed that separates the RNA and associated molecules from the outside world. A self-contained unit is formed—the cell.

These early cells were able to obtain their nutrients and energy directly from their environments. But that strategy can only go so far. The next step was for cells to start generating their own glucose, which could be used to generate and store energy to power their internal chemical processes. A major step forward in this process occurred with the invention of photosynthesis. This allowed early bacteria to use the energy from sunlight to convert carbon dioxide into glucose using hydrogen sulfide. In the first instance, this seemed ideal. Sunlight is omnipresent and easy to obtain; however, the appearance of a variant using water instead of hydrogen sulfide had dramatic effects. This invention occurred in the ancestor of current-day cyanobacteria, also known as blue-green algae. Cyanobacteria produce a byproduct, O_2, also known as oxygen. Over time, cyanobacteria's production of oxygen started to change the composition of the atmosphere. From being a rare molecule, oxygen became more and more abundant, eventually making up 10% of the atmosphere. This change took place between 2.4 and 2 billion years ago and is known as the Great Oxygenation Event.

The presence of so much oxygen in the atmosphere meant a dramatic change in how organisms metabolized fuel. Oxygen is a very reactive chemical. It can react with many organic compounds, including RNA and DNA. In effect, oxygen was toxic for much of the anaerobic life at the time. The Great Oxygenation Event might have been the Great Oxygenation Extinction Event. But as always in evolution, some life forms found a way, exploiting oxygen's reactive nature as a new way to provide energy with the development of oxidative metabolism. Oxygen and glucose are converted into carbon dioxide and water, releasing energy in the processes. Eventually, a balance developed between photosynthetic reactions in some organisms—mostly plants—that require carbon dioxide and produce oxygen, and aerobic reactions that require oxygen and produce carbon dioxide.

For anaerobic life forms, glucose is the main way to store and produce energy. But aerobic life forms do not produce glucose. They need to obtain it. You could argue that the birth of the aerobic life form was also the birth of a new type of life that needed to consume other life forms to obtain a vital nutrient. It was the birth of hunters. First, these were microbes eating other microbes. Quickly an evolutionary arms race started, with the preys finding ways to defend themselves and the hunters finding ways to hunt better. This was the beginning of the complex food webs we still see today, including those where some animals feed on plants and bigger animals feed on smaller animals.

It is difficult to determine when animals first evolved. Early animals didn't have shells or skeletons that easily preserve in the fossil record. Still, under exceptional circumstances, even soft tissue can be preserved in some types of rocks. Searching those rocks revealed no evidence for animals before 789 million years ago. Generally, the Ediacaran period (635–539 million years ago) is taken as the period from which animal fossils become more present. At the time, the earth was emerging from a long cold period when most of the land and oceans were covered in ice. As a result of the melting ice, sea levels rose, flooding large parts of the continents and creating shallow seas. In these shallow seas, the first animals appeared. These were simple, small, soft-bodied creatures creeping along the ocean floor, feeding on the cyanobacteria that were happily photosynthesizing away.

Somewhere around this time animals started to eat one another. This meant more purposeful movement, targeted at finding a prey. It also meant that some animals got better at avoiding being eaten. Chasing others and escaping from predators require coordinated movement. One major innovation in body plan that helped was that of bilateral symmetry. Bilaterians have a single line of symmetry running along their body, creating left and right mirror halves. As a result, they have a front "head" end and a back "tail" end, as well as top "back" and bottom "tummy" surfaces. Their movement is mostly forward with steering, rather than being able to move in all directions. Rather than impeding the bilaterians' flexibility, this gave them the ability to create dedicated muscles for powerful movement, directed by specialized sensors on the head. Nutrients can be ingested through the front, and waste safely expelled at the rear and forgotten about. Bilaterians represented a big step in the evolutionary arms race that had begun on the bottom of Earth's seas.

All of the time we have discussed so far—the first roughly 4 billion years of Earth's history—is collectively termed the Precambrian. This is in deference to the importance of the period that followed directly after. The Cambrian (538–485 million years ago) saw an explosion of diversity in animal life. In this period, the mineral calcium became more abundant in the water. Calcium is an important ingredient in teeth and in skeletons and armor. It gave both the hunters and the hunted the opportunity to experiment with new ways to build and protect their bodies. Animal life also moved away from the ocean floor. Whereas the Precambrian world was mostly two-dimensional, with animals foraging on the ocean floor, now animals moved in three dimensions. All this meant massive changes for animal life. To illustrate this, we will introduce our second animal.

At first sight, the lamprey might be taken for an eel. Just like an eel, it has an elongated body that slithers through water. But it is not. Eels are quite advanced vertebrates, while the lamprey is a survivor of a lineage of the most primitive vertebrates, the jawless fish. The lampreys that exist today are mostly parasites, with a round mouth opening full of small teeth that they use to attach themselves to fish. They then feed on the blood and flesh of their hosts, in some specialized cases even invading their internal organs. This behavior means lampreys are considered pests in some parts of the world, including in North America where they have invaded the Great Lakes.

In the Cambrian period, when vertebrates first appeared, the lamprey would have been a top predator. This period was a less peaceful time than the earlier vertebrate-less Precambrian. Like the sea squirt, the ancestors of vertebrates that lived during the Precambrian were probably filter feeders. In contrast, when the first vertebrates evolved during the Cambrian, they likely became more efficient at finding and ingesting food. Early vertebrates increased their body size, likely as a defense against predators, which in turn required more and better-quality food. They developed new ways of sensing the world, including paired eyes and an improved sense of smell, that provided enhanced sensing at a distance that might have allowed them to forage for food more efficiently. Rather than moving by pumping water through their body using ciliary movements, early lamprey-like vertebrates developed muscles for the purpose of movement. In all, early vertebrates had a substantially more active life than their ancestors and they had bodies to go with it. With these changes in lifestyle came innovations in the vertebrate brain.

Who has a brain

To understand early vertebrate brains, we'll look at commonalities between the lamprey brain and the brains of other vertebrates, including humans. It's safe to assume that any features that appear in both lamprey and human brains existed in our last common ancestor, which would have been a Cambrian vertebrate.

The lamprey brain, just like our own, contains cells specialized for communication exchange, the neurons. In a multicellular body like that of the lamprey, cells in different organs are specialized for different functions. Red blood cells transport oxygen, liver cells facilitate metabolism, skin cells protect the body, and so on. Neurons communicate.

Neurons send information from their cell body along a long fiber called an axon. They do this by regulating the relative abundance of different electrically charged molecules, ions, in and outside the cell. Certain ions are pumped between the inside and outside of the cell to maintain an electrical imbalance. Then when the cell wants to send a signal, ion channels open, and the electrical imbalance resolves itself, creating an electrical signal. This signal travels all the way along the axon. At the end of the axon it gives rise to many small branches before ending in synapses. The synapse forms the bridge between the axon of one neuron and dendrites of other neurons that carry information to the cell body. Communication across the synapse happens by means of release of specific molecules called neurotransmitters. This communication is therefore chemical. Dendrites, of which a neuron has many, take the signal from the synapses toward the neuron. Neurons work by receiving signals from many other neurons via their dendrites and then determining whether to fire off a signal along their axon themselves.

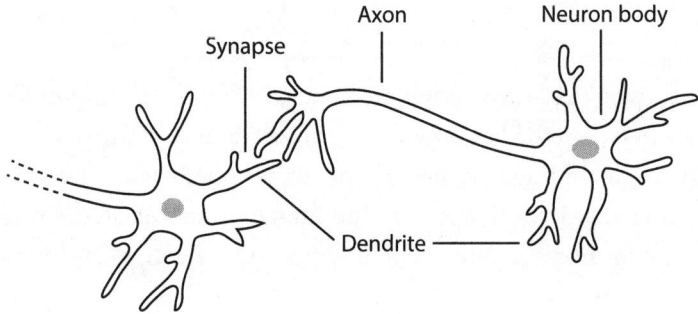

FIGURE 1.1. A neuron is a cell specialized for communication with other cells. An electrical signal is sent down the axon to a bridge with other neurons, called the synapse. From the synapses, dendrites carry information from many neurons to the cell body. If the receiving neuron fires in response to these incoming signals, it will send a signal down its own axon.

This ability to exchange information with one another through discrete signals sent along connections is the hallmark of neurons. It seems a pretty simple mechanism, but combine 86 billion of them and you've really got something.

Many of the genes that code for the creation of specific molecules involved in neuronal communication are ancient, some are even present in bacteria and therefore are thought to have evolved before the appearance of animals. Likely, these single-celled organisms' cell membranes contained voltage-gated channels that were important for regulating the ion balance of charged particles between the inside and outside of the cells. By controlling the opening and closing of channels in the cell membrane, these early cells helped maintain the right level of particles within the cell. The mechanisms for synaptic transmission were also present early. Some single-celled organisms that cluster together in colonies have mechanisms to release hormones into the water to communicate to one another. These two adaptations, control of electrical imbalance between the inside and outside of a cell and chemical transmission, form the basis of how neurons communicate.

Not all animals have neurons. Neurons probably evolved at least twice, once in the lineage leading to comb jellies and once in the lineage leading up to Cnidaria (anemones and jellyfish) and Bilateria (which include all vertebrates). Exactly when and why they appeared is hard to determine. The most likely scenario is that early animals benefited from electrically connected cells that could communicate the presence of a sensory stimulus outside the body to other cells. The first neurons, including chemical transmission at synapses, are thought to have appeared at a similar time as simple muscles did, helping to create a stimulus-detection-to-movement-execution pathway in the body. When animals started to engage in proper predator-prey arms races, muscles that allowed faster contractions appeared. Neurons guided the coordinated contraction and relaxation of these muscles. Once synaptic transmission was established, neurons could start forming connections with one another, allowing much more complex information processing.

Early in evolution, neurons were not clustered together to form a nice brain. They formed nets of nerves throughout the body. Only in some lineages of bilaterians did the neurons cluster together to form a brain. Something about the symmetric body plan and a clear front and back made this possible, but, since some bilaterians survive fine without such clustering, not inevitable. Brains seem to have evolved independently in arthropods (insects, spiders), mollusks (ranging from oysters and slugs to octopuses), worms, and chordates (the group including the sea squirt and vertebrates such as the lamprey). How brains evolved is different in each of those different groups; evolution does not always follow exactly the same path. For the purpose of this book, we will follow the path that includes our own brain, that of the vertebrates and represented by the lamprey.

In vertebrates, there is a central nervous system, with communication channels to and from the body traveling through the protected spinal cord to a central brain at the front of the body. The larger size of these animals put a premium on keeping different parts of the body coordinated, and their predatory behavior favored animals capable of making quick movements to obtain food, and to avoid becoming someone else's food. Both demands favored the evolution of a central system served by information provided from the senses by fast-conducting neurons. Indeed, this is when we see paired eyes appear as a highly specialized sensory extension of the nervous system.

The brain itself has the same basic subdivisions in all vertebrates. When we look at the brain of the lamprey at the start of this chapter, it contains most of the same subdivisions as our human brain. All vertebrates have what we call a hindbrain, a midbrain, and a forebrain. Within these subdivisions, we can identify some parts that have clear roles related to the active lifestyle of the vertebrate.

Of course, the primary thing you need to do to survive is, well, live. As discussed, the vertebrate body is large and complicated, with billions of different cells performing thousands of different functions. One important purpose of the brain is to coordinate all of these functions, like respiration, digestion, and metabolism. The first part of the brain we encounter when we move through the brain from the spinal cord is the hindbrain, and it is responsible for many functions related to keeping the basic body functions going. Present already in the ancestors of vertebrates, it is concerned with basics such as regulating breathing and sleep-wake cycles. Because of that last function, it is not hard to imagine this part of the brain has access to

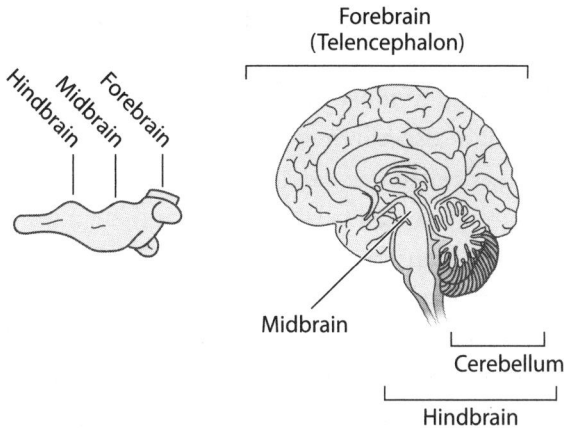

FIGURE 1.2. The main subdivisions of the vertebrate brain, shown on a lamprey brain and a human brain (view through the middle). Although the two brains look very different, the main subdivisions are the same.

information from the senses. Many of the basic responses to information from the senses are also mediated by this part of the brain.

What about movement, that function that seems so important in defining animal life? As a predator, the lamprey, like other vertebrates, needs to move a lot and that means it needs to process information in a different way than a static organism like the sea squirt. In a stationary environment, any movement you detect means that something else is moving. But if you yourself are tossing and turning, your sensory information might tell you that everything around you is moving even though it is perfectly still—it is you that is doing the moving. You need to be able to keep track of where all your body parts are at any given time. You also need to be able to detect quickly if your body does not respond in quite the way it usually does, for instance if your muscles are getting tired. To deal with these problems, the vertebrate brain contains an elaborate vestibular system.

The hindbrain of most vertebrates also contains a specific structure that deals with the ever-changing nature of the environment. It is called the cerebellum. It probably evolved in slightly more advanced vertebrates than the lamprey and is formed by an expansion of part of the hindbrain. The cerebellum has an extremely regular organization that makes it easy for neurons to exchange information with great temporal precision. This organization allows the cerebellum to make temporal predictions. If my brain sends motor commands to my muscles, the cerebellum predicts what kind of sensory feedback I should receive. It tells my brain what to expect if my arm moves as intended. This so-called forward model allows us to move much more accurately by anticipating the consequences of our commands, rather than just reacting to what we see. This is evident in patients whose cerebellum is damaged. Their movements tend to be jerky as if they are continuously overcompensating for small errors in their movement execution.

The regular cerebellar architecture does not only work for predicting the consequences of our own motor commands. If we follow a moving object such as a toy train as it goes through a tunnel, our cerebellum computes when it should emerge again. The cerebellum is often called the little brain, due to its smaller size in the human brain compared with our large forebrain (of which more later) and the fact that it is tucked away at the back of the brain. However, the cerebellum contains many more neurons than the rest of the brain. Cerebellar researchers sometimes display the brain upside down, with the cerebellum on top, just to emphasize its importance.

Movement is nothing without a good sense of the environment. Information from the senses gets linked in the brain to the control of the muscles that can effect some change in the environment. We might build a lot of brain in between to make this processing more complex, but that is the general idea. We can see that in the next structure, the midbrain. The midbrain is another structure that probably appeared in early vertebrates. In humans, it is located in the top part of the brain stem. It receives information from the senses and controls aspects of movement.

The midbrain's sensory-motor functions seem beneficial for a moving, predator-like lifestyle. The upper part of the midbrain receives information predominantly from the retina and to a lesser extent from other senses and organizes this in a topographic fashion. That means that information coming from neighboring parts of the environment goes to neighboring parts of the brain, forming a "map" of the outside world. The neurons in this part of the brain code how much a stimulus stands out from the background, forming a "salience map" of the external world. These neurons then project to parts of the midbrain involved in motor control. Mostly, these control the movement of the head or the eyes, allowing the animal to orient itself toward or away from nearby stimuli. When you see a frog reacting very quickly to the presence of a fly by launching its tongue, you see the midbrain in action.

This midbrain mechanism provides a very efficient mechanism for a foraging predator to react to stimuli in the near environment, but the early vertebrate predators also needed to process information about distant sources. This is where the forebrain comes in. Early vertebrates all lived under water and probably used olfaction, or smell, as the most dominant sense to tell them about the presence of prey at a great distance. However, smell diffuses with distance—the farther away you are, the less reliable the signal becomes. A simple strategy of moving up a smooth odor concentration gradient, from where a smell is less strong to where it is stronger, is therefore unlikely to be successful. The vertebrate predator needs to integrate information from different places over time to try to find out where the source of nutrition is. If the animal travels large distances, it might also need to keep a mental map to be able to navigate back to its home territory or find its fellow animals. In a word, the animal needs one or more types of memory. The telencephalon, a part of the forebrain unique to vertebrates, is predominantly concerned with these memory functions. Although the lamprey has a small and relatively simple telencephalon, in mammals and birds this part of the brain expands dramatically, with different parts evolving to accommodate different types of memory function required to deal with the foraging challenges encountered by each species. Much of this book will therefore discuss differences in telencephalon organization between species.

———

Becoming a predictive forager

In the previous two sections, we used the examples of the sea squirt and lamprey to gain a better understanding of how the vertebrate brain evolved. We reconstructed animal life up until the appearance of vertebrates and saw that different kinds of brains evolved in a couple of lineages of complex animals, including the lineages leading to vertebrates. The different subdivisions of the vertebrate brain—hindbrain, midbrain, and forebrain—are consistent with the idea that the brain evolved to help organisms become better foragers, but this evidence is largely circumstantial. In order to really prove that the brain is a foraging device, we will need to identify a general character-istic that is specific to brains and essential to foraging.

As we stated above, the key problem to solve in foraging is that you need to obtain nutrients that are sparsely distributed in space and time. To obtain the nutrients, and to do so in a way that costs less energy than it takes to obtain them, it is not suf-ficient to be merely very good at detecting food and going after it. A wolf that can only find deer and then run toward them will mainly go hungry. It's too easy for deer to find a counter strategy—in this case, very good early detection of the predator and then high-speed flight. What the successful forager needs to be able to do is to anticipate its prey's actions. If the hunter knows that at a certain time of the day a mother deer will take her young to drink at the water outside the cover of the trees, they can use that information to their advantage. A good forager needs to be capable of *prediction*. That, it turns out, is a very ancient capabil-ity of neurons. To illustrate this, we introduce our third animal.

Bees are winged insects, members of the phylum of arthro-pods that also includes arachnids and crustaceans. It has been known for a long time that bees can display very complex be-havior. The 1973 Nobel prize in physiology or medicine was given to three pioneers in the study of animal behavior. Of these three, Niko Tinbergen and Konrad Lorenz are the best known, but Karl von Frisch's work on bees has become almost iconic. He discovered that honeybees can communicate the location of flowers containing nectar to one another. Bees of course do not speak. Instead, they use a dance. By tailoring their dance, they can indicate the location of the flowers relative to the angle of the sun and their distance from the nest to other bees. The press release announcing this Nobel prize to animal research-ers, rather than medical doctors, emphasized that many of the mechanisms guiding the behavior of these animals are likely to be seen in humans as well, justifying an award in the field of medicine. How true that turned out to be.

As we discussed in the previous section, arthropod brains evolved separately from those in the vertebrate lineage. Consequently, the arthropod brains looks very different from the vertebrate brain. But some of its most basic foraging apparatus is similar to that of the sea slug and the lamprey.

This seemingly universal foraging mechanism was studied quite extensively by neuroscientists and biologists, but also by engineers. A group of engineers, led by Read Montague, spent a substantial part of the early 1990s thinking about the foraging behavior of the bee. Experimental studies had shown that bees can learn to distinguish flowers that give more nectar than others based on their color. In a typical experiment, bees were confronted with flowers of two different colors, of which one was associated with on average a higher amount of nectar than the other. After a few exposures, the bees were able to choose in such a way as to maximize their expected return. To be able to do this, the engineers realized, it is not enough for the bees to be able to detect the reward. They need to form an association between the flower and the reward. Put simply, in their brain, their module detecting the color of the flower needs to be wired up in a specific way with the module that detects the nectar.

A possible mechanism for neurons to learn such associations had been proposed before. In 1948, Donald Hebb published his book *The Organization of Behavior*. In this book he proposed that if two neurons fire at the same time, the synapse between them becomes stronger. That means that the next time one of those neurons fires, it becomes more likely that the other fires as well. This is popularly summarized as "fire together wire together."

At the time, Hebb had no physiological evidence for this mechanism, but his suspicions proved accurate, and today the mechanism is known as "Hebbian learning."

Hebbian learning is a very simple but powerful way for associations to form. But it also has limitations. Often, a stimulus predicting something nice like nectar doesn't occur at the same time as the nectar. It is lovely for the bee that it can detect that it has received nectar once it has landed on a blue flower, but what would be more useful for foraging is that the bee can predict in advance that the blue flower will yield more nectar. Moreover, the system needs to be able to deal with unreliable cues. What if the blue flower had been visited earlier by another bee and therefore has no more nectar? The bee should not learn to never again try a blue flower, but simply to adjust its expectations a little. How can the brain of a foraging bee deal with those challenges?

Let's deal with the second problem first, the unreliability of the information. The solution is not for the brain to just learn lots and lots of associations, but to learn to compute the value of a given situation, an abstract measure that indicates goodness. Let's assume, for instance, that a blue flower always gives a little bit of nectar, but a yellow flower gives much more nectar but only part of the time. In nature, bees learn to balance this trade-off. They generally prefer a reliable flower, but one that gives particularly high amounts of nectar part of the time might just tip the balance. Read Montague and his engineer colleagues tried to build a model of how bees solve this dilemma.

When Montague and colleagues were researching bees, a type of neuron that seemed ideal to help with the problem of predictive foraging had just been discovered in the bee brain. Stimulating this type of neuron made the bee behave as if it had just been rewarded with nectar. In other words, firing of this neuron presented something that felt to the bee like "goodness." Building a computational model of this neuron's behavior seemed a promising place to start. Montague and his colleagues assumed that the neuron received some information about the color of the flower the animal encountered and that the neuron weighted that with the inherent goodness of the flower. The neuron also received information about the actual nectar the animal received. The neuron's firing was taken to be the difference between what it expected to receive based on the flower and the nectar it actually received.

In other words, the neuron did not simply indicate when something was good, but when something was better than expected. It created a type of "teaching signal." That teaching signal could then be used to learn about the goodness of the flowers by changing the weighting of goodness associated with that flower. If a flower reliably led to reward, its weights would be adjusted to reflect a high goodness. If a flower only occasionally led to reward, its weights would be adjusted to reflect that fact. Now, when the bee encountered a flower, it could use the information stored in weights, to guide its behavior. Sure enough, an artificial model of these neurons learned to favor reliable flowers but to sample the less reliable flower if its occasional nectar yield was big enough. By coding for an abstract measure of reward, rather than only actual reward, that is, nectar, the system was able to learn subtle balances in the environment.

Assigning an abstract reward value to the flowers and coding differences between expected and actual reward solves one of our foraging problems. What about the other one, that a forager needs to be able to learn about predicting rewards way before they occur? At the same time that Montague was thinking about bees, Wolfgang Schultz at the University of Cambridge was performing similar experiments in an animal much closer to humans, the macaque monkey. Schultz's monkeys performed a simple task. They received a visual stimulus—a light—that indicated to them to make one of two possible movements with a lever. When they make the correct movement, they receive a juice reward. At the beginning of the experiment, the monkeys were just guessing which was the correct response, but when they hit the correct lever, a juice reward was delivered.

While the monkeys were performing these tasks, Schultz and his team recorded from a particular type of neuron in their brains, termed dopamine neurons. The juice reward elicits strong firing of the dopamine neurons. After a while, the monkeys became proficient at the task and always selected the correct lever. Once the monkey had become proficient at the task, receiving juice didn't have the same effect on the neurons. Now, when the monkey received a reward, the dopamine neuron was silent. Instead of firing when the monkey received the juice, the neuron actually fired as soon as the instruction light came on.

How did this happen? Once the monkey has learned the pressing the level reliably results in juice, the juice is no longer an unexpected reward. Rather, it is the sudden appearance of the instruction light—ah, I get to press the button and get a reward—that is the first sign that something good is about to happen. Just as in the case of the bee, the dopamine neuron signaled not the reward the animal received, but rather a sudden change of fortune. What was new in the Schultz experiment was that the signal traveled in time, from the moment of the reward, to the moment of the instruction.

The neuron Montague and colleagues modeled in the bee is evolutionarily related to the dopamine neuron in monkeys. So let's look at our foraging bee again and see if the finding of Schultz in monkeys can be of use in understanding the similar neuron in bees. Imagine the bee is looking for nectar and encounters a blue flower. The dopamine system, or its bee equivalent, is just firing at baseline level. The bee lands on the flower and encounters nectar. The dopamine neurons start firing away, this is much better than expected. The teaching signal is used to train the system.

After many such encounters, the animal has learned the predictive value of the blue flower. Now, when the bee is flying around minding its own business, the sight of a blue flower is an unexpected joy, and the dopamine neurons fire. When it lands on the flower and receives the nectar, however, all is as expected and the dopamine neurons are back at baseline.

Now, imagine what happens when the bee unexpectedly does not receive nectar from the blue flower. The dopamine neurons reduce their firing rate below baseline. The teaching signal is now going the opposite way, broadcasting the information that all is less good than expected.

Montague and colleagues proposed that these simple changes in dopamine firing provide exactly the information necessary to learn about the predictors of reward. But the dopamine system does more. Dopamine neurons are located in a small part of the midbrain in vertebrates, but project to a wide range of brain areas, including a large part of the telencephalon, the part of the brain that shows such variability in size and organization across different foraging vertebrates. At these faraway areas they release the substance dopamine, which helps speed up the synaptic strengthening we see in Hebbian learning. As one dopamine researcher remarked, dopamine neurons don't say much—just a simple assessment of "better than expected," "as expected," or "worse than expected"—but what they say gets heard in a lot of places in the brain. Their simple teaching signal can be used to train much more complicated associations and relations in the telencephalon, the part of the

Before learning

After learning

After learning — no reward

→ Time

FIGURE 1.3. Dopamine neurons and the foraging bee. Before any learning (top row), the neurons increase their firing (indicated by the stacks of black boxes) when the bee receives nectar. After learning (middle row), the neurons increase their firing already on encountering the flower, since that predicts the nectar. They are silent when the nectar, which is now fully predicted, occurs. However, if the nectar does not arrive (bottom row), the neurons decrease their firing. Together, the firing pattern of the dopamine neurons presents an excellent teaching signal.

brain involved in the various types of memory many vertebrates use to aid their foraging. The dopamine system, and its evolutionary cousin in the bee, constitute a simple, generic, and flexible system for learning what environmental situations, actions, internal body states, or memories are likely to lead to good outcomes. Exactly what we would need for our reward-seeking forager.

Recap: The sea squirt, the lamprey, and the bee

We started this chapter with a dilemma. Brains are generally relatively large, metabolically expensive, and fragile organs to have. What benefits do they provide to outweigh those disadvantages? The answer is that the brain offers the organism a solution to complex foraging problems. As animals became bigger and started to live a more active lifestyle, their requirements for nutrients became bigger and the methods to obtain them became more varied and more complex. Early vertebrates already possessed a centralized nervous system, packing together brain cells capable of rapidly communicating with one another through a combination of electrical and chemical signals. New parts of these brains dealt with the integration of sensory and internal systems and steering muscles to affect the environment. An ancient learning system provided a way to combine different sources of information, creating a predictive, forward-looking foraging system.

Although the foraging life of the lamprey is complicated, as vertebrates go it has a relatively simple life. Other branches of the vertebrate tree have to deal with much more complex situations, requiring them to deal with faster changes in their environment and more diverse and unreliable information. Over the course of evolution, many distinct lines of vertebrates have faced many different environments. Some animals were able to adapt, as their brains adjusted to solve their new problems. Those that didn't died off. In the next chapter, we'll explore a major shift in vertebrate brain evolution, which happened about 380 million years ago, when some vertebrates left the water for a life on the land.

2

The dragon and the shrew

MAMMALIAN NEOCORTEX

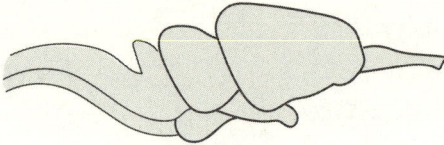

Komodo Dragon   Common shrew

In *The Times* of June 27th, 1927, Dr. Peter Chalmers Mitchell, Fellow of the Royal Society and Secretary of the Zoological Society, reports that the new reptile house of the London Zoo is to be opened to Fellows and visitors at 3:30 that day. He starts with a description of the beautifully designed exterior of the building, followed by a description of the lobby and the exhibition hall, which has a central island block surrounded by wide corridors for the visitors. He assures the reader the poisonous snakes and dangerous constrictors are confined to the central island, ensuring "complete safety" for the visitors. He goes on to describe the various animals on display and, in quite some detail, the many exquisite design features of the reptile house, including "electric heaters embedded cunningly on shelves of rockwork [on which] special beams of electric light play." Finally, he gets to the "best kept secret" of the new reptile house: "Through the kindness of Dr. Malcolm, a Fellow of the Society long resident in the East, the Dutch authorities have been induced to send to the society two living examples of the famous 'dragons' of Komodo [. . .] the first living examples to reach Europe." Dragons in London, that merited a report in *The Times*!

In the nineteenth and early twentieth century, large parts of the Malay Archipelago were occupied by the European colonial powers of the day. As history teaches us, the colonial powers were primarily interested in exploiting the natural resources of the archipelago for purely economic reasons. Some individuals, however, were also interested in the natural history of these lands. The Malay Archipelago was where the British naturalist Alfred Russel Wallace traveled between 1854 and 1862 to collect specimens to sell to the upper class gentlemen collectors in his home country. His observations led him to speculate on a theory of evolution by natural selection, in parallel with Charles Darwin. It was also here that Eugene Dubois found the leg bone and skull cap of *Homo erectus*.

To promote the exploration of natural history, the Dutch colonial government established a zoological museum in Buitenzorg (now Bagor) on the island of Java, near the summer residence of the Dutch governor-general. In 1910, the director of this museum, Peter Ouwens, received word from a lieutenant Van Steyn van Hensbroek, who had heard from the local inhabitants about a land crocodile living on the islands of Flores and Komodo. Van Steyn van Hensbroek managed to send a skin and photograph of the animal to Ouwens before being transferred to a different post. He had tried to capture more, but the local inhabitants refused to help him, as the animals bit and kept people "at a respectful distance by powerful blows with their tails." Ouwens then sent a nameless "native collector" of the museum to obtain live specimens for him to study. Armed with dogs and local help he managed to bring home two large cadavers and two live young specimens. Ouwens communicated the existence of the animal to the European scientific world in the *Bulletin du Jardin botanique de Buitenzorg* in 1912, labeling it *Varanus Komodiensis*. Although the giant lizard looks a tad dopey, it is a ferocious predator that doesn't hesitate to attack wild boar, deer, goat, horses, and on occasion humans. Its victims typically bleed to death due to the anticoagulant in its saliva. Often, the animal will simply bite its prey and wait for it to die on its own before eating it. Reports of subsequent expeditions to capture and study the animal soon entered popular culture, together with a name that better appealed to our imagination: the Komodo dragon. No wonder the London Zoo considered them the prime exhibit of their new reptile house.

In the previous chapter we discussed the vertebrate brain. The lamprey is an example of a vertebrate that lives its entire life in the sea. Reptiles and mammals form the two branches of the vertebrate family tree whose ancestor made the transition from the ocean to a permanent life on dry land. The Komodo dragon is now recognized as one of the largest living reptiles and, conveniently for us, has one of the largest reptile brains to study. The dragon's brain, just like a mammalian brain like our own, has all the vertebrate subdivisions we outlined before. When we put the dragon brain and that of a mammal side by side in this chapter's opening figure, however, they look very different.

The dragon brain has two bulbs at the front, connected to the rest of the brain via two long nerve pathways. These bulbs are the olfactory bulbs, a part of the forebrain involved in smell.

The mammalian brain has these as well, but they are very small in comparison to the rest of the brain. This suggests that the dragon relies much more on smell than mammals do. Indeed, Ouwens' first report suggests the dragon is completely deaf—"if only care is taken, that the animal does not see the hunter, the latter may make as much noise as he pleases, without the animal being aware of his presence." In the mammal, the small olfactory bulbs are almost covered by a large structure—this is the neocortex, the topic of this chapter. At first glance, the mammalian brain looks like the reptile brain with the neocortex as a large new part put directly on top of it. This is precisely what early evolutionary neuroscientists thought.

One of the most popular ideas about brain evolution is that the big human brain is superior to all other brains because it basically consists of a primitive reptile brain with more advanced parts on top. In popular culture, when we think with our "reptilian brain" it means we let ourselves be guided too much by our primitive urges, but hopefully then we come to our senses and let our "rational brain" take over. In scientific terms, this was referred to as the triune theory of brain organization. For a long time this was one of the most advanced theories on the evolution of the human brain. The theory in its most complete form was proposed by the physician Paul MacLean. He compared the brains of different animals under a microscope and concluded that certain parts of the human brain were the same as those of lizards, some parts were the same as those of simple mammals but not present in lizards, and some were unique to the human.

MacLean proposed that the human brain consists of three parts: the R-complex that evolved first in reptiles and is concerned with basic survival behavior such as aggression, territorial behavior, and ritual displays; the paleomammalian complex or limbic system that evolved first in early mammals that is concerned with motivation and emotion involving feeding, reproduction, and parental behavior; and the neocortex that evolved in modern mammals that is involved in higher behaviors such as cognition and abstraction. These three parts are colloquially known as the "lizard brain," the "emotional brain," and the "rational brain." The higher regions supposedly evolved to temper the behaviors of the lower regions. This fits well with the popular notion that we humans usually behave rationally, but sometimes our evolutionary more primitive lizard brain takes over and leads us on a path of aggression and indulgence. In essence, the theory says, we all have a little dragon in our brain.

This triune theory of the brain is, however, wrong. We are mammals, and mammalian ancestors and reptile ancestors diverged from a common ancestor about 320 million years ago. Over the course of those 320 million years, reptiles and mammals both modified their brains. The predator lifestyle and almost cartoonish evil look make the Komodo dragon seem like a monster from a bygone age, and having part of its brain controlling our more primitive urges is a good party story. But we don't have a little dragon brain in our head. The real story, though, is no less fascinating. It is a story of how animals invaded and conquered the land and a story of two mass extinctions. It is also the story of how mammals developed this unique piece of brain anatomy called the neocortex.

A tale of two extinctions

Three hundred and eighty-five million years ago, the world was very different from today. The supercontinent of Gondwana covered much of the Southern Hemisphere, with the other continents of Siberia and Laurussia in the Northern Hemisphere and over the equator. Over the next 50 million years, Gondwana moved slowly toward the other continents to form what would eventually become the single supercontinent of Pangea. Most of the land had a temperate climate. Sea levels were high, with much of the land covered by shallow seas.

In this world, some vertebrates made the move from an aquatic existence, such as that of the lamprey, to living on the land. We aren't sure exactly how this happened, but the prevailing theory is that frequent droughts might have led to the formation of pools, forcing certain aquatic organisms to travel over land to move between bodies of water. High oxygen levels and the absence of predators might have made exploring the land a viable option. On land, these vertebrates encountered the pioneers that had already made the transition to land earlier—plants and invertebrate animals such as insects, worms, snails, and relatives of spiders. As with many evolutionary pioneers, the early vertebrates exploring the land were likely predators interested in these invertebrates.

Living on the land and the predatory lifestyle required extensive modifications to the vertebrate body. The mechanics of movement without the buoying effect of water meant fins needed to change into limbs capable of carrying the weight of the body. Just as Eugene Dubois searched for a missing link between humans and great apes, other researchers have tried to find species intermediate between fish fully adapted to life in water and animals fully adapted to the land. One of them is the paleontologist Neil Shubin. Where Dubois had reasoned, fortuitously it turned out, that his missing link should be in Asia, Shubin reasoned that fossils of animals making the water-to-land transition should be located in rocks left over from the Devonian period, about 375 million years ago. Anything older than that and all fossils still look like fish, more recent than that and fully formed land animals are present. They found rocks of the appropriate age in the Arctic regions of Canada and set to work finding a missing link between aquatic and land-dwelling vertebrates. Shubin and his colleagues spent four expeditions over six years in this rather inhospitable climate searching for their missing link. But on the last day, when they were ready to give up, they found their fossil.

Tiktaalik, as it is now known, is a transitional species between a fish and a tetrapod or "four-footer." It probably lived in water most of the time but was capable of excursions on land, as we can see from its limbs: its fin bones already resemble the sturdier bones and movable joints of land vertebrates, allowing it to prop itself up. Tiktaalik and its fellow vertebrate pioneers had to adapt to the challenge of living on land in other ways as well. The way land predators bite, with vertical jaw motion, is very different from the way water predators, like the lamprey, bite, which largely uses suction. The shift from suction biting to vertical biting led to elaborate changes to the head. Unlike fishes, Tiktaalik had a neck that allowed it to move its head independently from the rest of the body. Breathing, of course, in land animals is no longer done through gills but through lungs, another thing the early land pioneers needed to address. Even the sensory apparatus needed to change. Some fish use electroreceptors to detect objects or prey in the environment, but these sensory organs work only in water and would be useless on land. Sound travels much less far in air than in water, and eyes outside of water need to adapt to a different refractive index and be kept from drying out. All these body modifications can be seen in early land vertebrates as they evolved to live full-time outside the water.

One major problem left unsolved by the early land vertebrates was that of reproduction. For this, the animals always had to return to the water where they laid their eggs. Eggs laid on land were unable to prevent their fluid from evaporation, meaning the embryo would dry out and die. The reproductive cycle of early land-based vertebrates resembled that of today's amphibians, returning to the water for a part of their life cycle. Their offspring likely went through a tadpole-like stage when they lived in water, before a metamorphosis to a land animal.

By 312 million years ago, a solution appeared. This solution was a membrane, the amnion, around the embryo. The amnion fills with fluid that provides a protective environment for the embryo to keep it from drying out. A complex system of gas and fluid exchanges allows the embryo to breathe, obtain nutrition, and excrete waste. This enabled the animals to lay their eggs on land or later, in some branches of the vertebrate tree, to retain them inside the mother. The group of animals adopting this solution are called the amniotes, after this membrane. Free from the necessity of staying close to the water, these animals were free to explore new niches on the land, ranging from the mountains to the deserts, and from the treetops to the underground. Amniotes quickly diversified and became the dominant land animals on Pangea. They split into two lineages: the Synapsida and the Sauropsida. The Synapsida are those amniotes most closely related to modern mammals. The Sauropsida are those amniotes most closely related to modern reptiles, including birds. (Yes, evolutionarily speaking, birds are a branch of reptiles, as we will discuss in chapter 7.)

Although we commonly consider our current time as "the age of mammals," the mammalian ancestors between 300 and 250 million years ago could justifiably have termed their time an age of "mammalian ancestor dominance." Many of the evolutionary adaptations we have come to associate with mammals first appeared in this period, in a group of synapsids we call the therapsids. Therapsid innovations include changes to the skeleton to allow a mammalian way of walking with the legs underneath instead of beside the body, warm-bloodedness, perhaps fur, and the specialization of different types of teeth in the mouth such as incisors, canines, and molars. The guts of some of these animals changed, providing a home to microorganisms that are essential for breaking down plant matter. This was the birth of dedicated herbivores and the start of a stable ecosystem consisting of many herbivores feeding on plants and a few carnivores feeding on the herbivores. The niche of large land animals was dominated by therapsids, and their life seemed good. But things were about to change.

Two hundred and fifty-two million years ago, life nearly ended on Earth. As mass extinctions go, the end-Permian extinction was the big one. Seventy percent of terrestrial vertebrate species were wiped out, as well as more than 80% of marine species. Every form of life was affected—plants, animals, terrestrial, marine, vertebrate, invertebrate. Until the extinction we are currently living through, it was the only one to significantly affect insects. In fact, the end-Permian extinction might not have been a single event, but a sequence of successive events. Its causes remain a topic of speculation, but it seems likely that sustained volcano eruptions in Siberia set in motion a series of catastrophic events. The results of the eruptions can still be seen in the landscape as the Siberian traps, a region of about seven million square kilometers covered by volcanic basalt rock. Although the basalt lava from the eruptions was devastating only for the direct environment, what the eruptions released into the atmosphere had worldwide consequences.

Repeated spewing of large quantities of carbon dioxide led to a sustained global warming. This triggered other events, including the release of methane locked in the sediments of the polar sea. Such positive feedback mechanisms led to runaway global warming. Atmospheric oxygen levels plummeted. Sulphur dioxide and chlorine gas from the eruptions led to acid rain that devastated plant life, which in turn led to massive soil erosion. Ecosystems everywhere collapsed.

After the end-Permian extinction, the world in many ways may have resembled the earlier conditions of low oxygen and high temperatures known from the Cambrian Period 540 million years ago, when vertebrates first became established in the sea. Animal life took millions of years to recover and was probably hit by repeated setbacks. But when life does recover after an extinction event, there are many empty niches for the survivors to fill. Evolution gets a chance to experiment with new innovations. Those animals that did survive the mass extinction were probably generalists, able to survive in a broader range of conditions than the extreme specialists.

In the first couple of million years after the end-Permian extinction, a family of species from the synapsid lineage became the dominant animal on land. They were called *Lystrosaurus* and seemed to have been a piglike herbivore that didn't care much what it had for dinner. Not being picky, it had the best chance to survive on what was left. The members of this family probably ranged from cat sized to cow sized, taking over the niche of large herbivores for a while. Other niches, such as that of large carnivores and smaller predators, remained unoccupied. After a while, new animals appeared to claim the unoccupied niches. Instead of the synapsids, it was the reptiles that got the upper hand. First, relatives of crocodiles became the unlikely pretenders, evolving into a range of species, including large herbivores, slender greyhound-like runners, and large marine predators. At the end of the Triassic period, about 201 million years ago, another smaller extinction event shook up the reptile kingdom and the quintessential reptile took over. The age of the dinosaurs had begun.

With the dinosaurs taking over the position as large land animals, little was left for the mammalian ancestors. They had one opportunity, though, that was unavailable to the reptiles. Unable to compete with the large reptiles that roamed the Earth in the daytime, mammals opted for the leftover niche of small, nocturnal animals. In that niche, pre-mammals and early mammals started experimenting and diversifying. The jaw of mammals changed, with some of its bones becoming part of the hearing system. This way, mammals evolved a highly specialized sensory apparatus helpful in the nocturnal niche. Their vision also adapted, with mammalian eyes showing an abundance of light-sensitive rods on their retina.

The early mammals were able to pursue the night life thanks to the fast metabolism and warm-bloodedness they inherited from their ancestors. This allowed them to become small and be active in the night's cold while reptiles always required daytime warmth to be active. It came at price, though, as maintaining body temperature requires high-quality, energy-rich food. One type of food that fulfills these criteria is insects. Most likely, the earliest mammal was a nocturnal insectivore, similar in size to modern day shrews. While dinosaurs and their relatives roamed the day proclaiming the "age of reptiles," mammals were optimizing their niche adaptation and awaiting their chance.

The age of the reptiles came to an abrupt end. Sixty-six million years ago marks another major extinction event. It was not as big as the end-Permian extinction, but enough to kill off most of the dinosaurs and put an end to large reptiles as the dominant land animal. In fact, it is likely that no animal bigger than 25 kilograms survived. The cause of this extinction has long been disputed. I am reading my old children's books about dinosaurs to my children now and they still discuss a variety of wild ideas as to why dinosaurs suddenly disappeared. Climate change, volcanic eruptions, little mammals eating dinosaur eggs, radiation due to the supernova of a distant star—are all suggested. The truth was spectacular.

In the 1980s the father and son scientists Luis and Walter Alvarez started pointing to the high levels of the metal iridium detected in geological layers from the time of the extinction event. Iridium is much more present in asteroids than in the Earth's crust. Luis and Walter suggested that a major impact event was the cause of the extinction of the dinosaurs. This theory was long disputed. Geology at the time was a science that didn't like to deal with sudden biblical-type catastrophes. Slow and gradual change was the mantra. But in the early 1990s a large crater buried underneath the Yucatán Peninsula in Mexico was identified. At 180 kilometers in width and 25 kilometers in depth, it is one of the largest impact structures on Earth. When the asteroid hit, the impact released energy equal to a billion times the atomic bombs dropped on Hiroshima and Nagasaki. A dust cloud blocked out the sun for a year, leading to the collapse of photosynthesis in plants and setting off a larger collapse of many ecosystems. For the specialized dinosaurs, this was the end of the road.

Devastating as the results of the impact winter were, after every extinction event there are survivors. For some mammals, already adapted to living in cold, dark niches where food was not abundant, this was their chance. Evolution again started experimenting and filling niches that had become vacant. This time, the niche of large land animals was occupied by mammals. This was the start of their age.

———

Finding the neocortex

Reptiles and mammals are the results of the evolutionary adaptations by their ancestors in response to a variety of challenges over the course of their evolutionary history. As we have seen, their different solutions are reflected in their different bodies and lifestyles. But how do the brains of reptiles and mammals differ?

If we look at the figure of the shrew's brain at the beginning of the chapter as a mammal example, we see that its brain is covered by a mantle-like structure. This is the neocortex that Paul MacLean identified as the uniquely mammalian rational brain. The neocortex is part of the pallium, the upper part of the telencephalon and a part of the brain that is present in both reptiles and mammals. As we saw in the previous chapter, the telencephalon is the part of the vertebrate brain where many of the changes related to different foraging solutions can be found. The mammalian and reptilian pallium are a prime example of a part of the brain where such changes occurred. By studying animals from both lineages, we can compare the structure, type, and distribution of neurons in different parts of the pallium and see if they look alike across brains. This would be a first criterion of establishing whether the parts of the brains are "the same" or "different" between reptiles and mammals.

However, just looking at the finished product, as Paul MacLean did, can be misleading. This is easily demonstrated outside the brain. For example, the fins of many fishes look very different from our limbs. They certainly work in a different way. But if we look at how they grow during development, we see that our limbs and the fins of fishes are actually quite similar. They are even under the control of the same genes. Therefore, besides comparing the finished products, comparative scientists often also study the development of biological structures. We can do the same for the pallium of the brain. We can study its development and see if different parts of the pallium follow a similar trajectory in different species, even if the end result might look a bit different. If we find that areas in two brains follow the same pattern during development and rely on the same genes, we call them *homologous*. This means they have the same origin, even if their exact look or function has changed.

Let's see how the pallium of the Komodo dragon compares with that of a small mammal such as the shrew. If we cut through the pallium from left to right and look at the resulting section, we see a slice of brain tissue with some clearly different subsections. In the reptile brain, toward the top and the middle part of the brain is a large space normally filled with cerebrospinal fluid that provides mechanical and immunological protection for the brain. Even more to the middle is a layer of brain cells that is termed, because of its position, the medial cortex. If we look at how this part develops and which genes are expressed here, we can find a very similar structure in the mammalian pallium, but there it is called the hippocampus. The mammalian hippocampus is a structure consisting of three layers of neurons that are very important in spatial navigation. For a foraging land animal, understanding your environment is important, and the hippocampus is able to link places together into a mental map. In fact, this function is so useful that it might have formed the basis for a whole range of memory functions. We will explore that story in chapter 7. For now, it is important that the medial

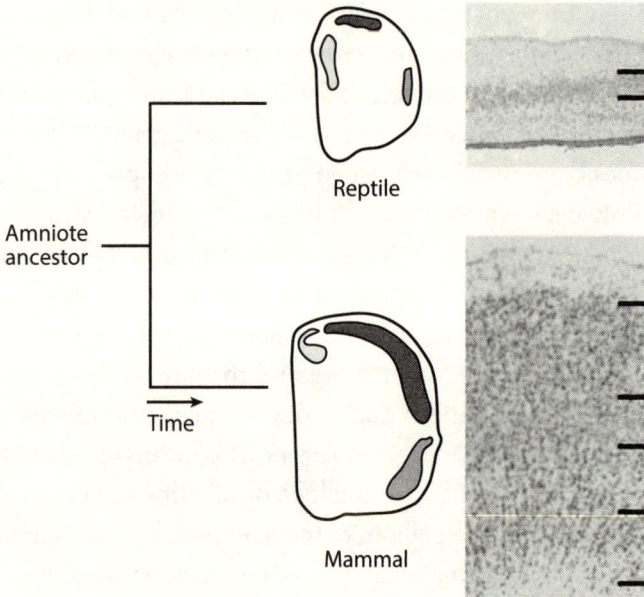

FIGURE 2.1. A slice through the brains of a reptile and a mammal show that their brains have similar areas, even though their size and location are a bit different. Each has a medial cortex or hippocampus (lightest gray), lateral cortex (medium gray), and dorsal cortex (darkest gray). The dorsal cortex of the reptile (Komodo dragon) is similar to the dorsal cortex or neocortex in the mammal (shrew). The neocortex is organized in a much more complex way, as can be seen on the right. Its neurons and their connections are organized in a six-layered structure allowing complex information exchange, whereas the reptilian dorsal cortex has fewer layers.

cortex or hippocampus is present in similar form in both reptiles and mammals and, therefore, was likely part of their amniote ancestor.

Toward the other side of the reptilian pallium, near the outer border of the brain, we see another structure. Because it is located at the side, it is called the lateral cortex. Again, if we look at how this part of the brain develops and which genes are important in its formation, a similar structure can be found in the mammalian brain. In mammals, it is called the piriform cortex, important for the analysis of smell. Smell is used by both reptiles and, to a lesser extent, mammals. That it is much more important to reptiles than to mammals is evident by the size of this part.

In the mammal, the piriform cortex takes up a lot less space compared with other parts of the pallium than the homologous lateral cortex does in the reptile. The same can be said, as we saw earlier, for another part of the brain important for processing smell: the olfactory bulbs. These primary smell processors stick out from the front of the brain, such that they are located close to the nasal cavity. They are more prominent in the Komodo dragon but are barely visible in the mammal because they are covered up by the rest of the brain. This fits with what we established above, that early mammals started relying on other senses apart from smell, including vision and hearing.

The final part of the brain we see in our cross section looks very different between the dragon and shrew brains. In the dragon, what is left is a small part consisting mostly of a single layer of large neurons with some smaller inhibitory neurons interspersed. This part of the brain receives input from the outside world from other parts of the brain and provides output back. Most of its input is visual information. This small, inoffensive part of the reptile brain is called the dorsal cortex, because of its location high up, or dorsal, in the pallium. In mammals, the picture is very different. The remaining part is much larger—it can take up most of the space in the skull for some mammals—and it is called the neocortex. Rather than a single layer of neurons, this part of the cortex in mammals shows a complex organization in up to six distinct layers. Different types of neurons are located in different layers. Inputs from other parts of the brain enter in one layer, the information gets processed through a series of different neurons in the different layers before it gets transmitted outward from another layer.

While the reptilian dorsal cortex is just a thin layer, the mammalian neocortex is organized into a column of neurons across six different layers, which allows for much more complex processing of information between the moments when information comes in and when it goes out. Because of the column, this part of the mammalian brain is much thicker than the reptilian dorsal cortex. It looks very different from the dorsal cortex, but is homologous to it—the same structure in both reptiles and mammals, not one evolved from the other. All mammal brains contain a neocortex, honed over millions of years of evolution. We don't know for sure if the reptile dorsal cortex is anything like the dorsal cortex of the amniote ancestor of both reptiles and mammals, but based on the differences in complexity between the reptilian dorsal cortex and the mammalian neocortex, it is safe to say reptiles did not spend their evolutionary capital upgrading this part of the brain, whereas mammals did.

The neocortex provides mammals with an extremely flexible part of the brain, allowing for increased complexity in the way mammals respond to the world around them. We can see the neocortex as a two-dimensional sheet of columns of neurons. The cortical column provides a handy processing unit. The neocortex can easily be upgraded by building more of them without fundamentally having to change what is already there. By adding cortical columns, this sheet can become very large in some mammals, including elephants, dolphins, and primates such as ourselves.

When the two-dimensional sheet that makes up the neocortex becomes very large, nature's way to store it inside our skull is to fold it. That is why our brain looks crinkly and has the large grooves that you see when you look at it from the outside.

If you unfolded the wrinkly neocortex and laid it out, you would see that not all parts of the neocortical sheet are exactly the same. Specializations have occurred. Some parts receive lots of information from other parts of the brain, with some specializing in visual information, some in auditory information, and some in sensations from the body such as touch or temperature. Other parts provide output to the rest of the nervous system, for instance the "motor cortex" that helps control movement. These areas dealing with one specific type of information are often called "unimodal cortex."

In between the unimodal parts of the neocortex are areas that generally do not interact much with the rest of the brain at all. They communicate only with the unimodal areas and each other. They are called the association cortex. Just as the different layers of the neocortex added computational complexity, so does the association cortex. By building extra processing units devoted purely to reprocessing the information already present in the neocortex, the computations can become increasingly complex. For instance, different species' neocortexes are capable of combining information from different senses—that is, from different parts of the unimodal areas—to different extents. All in all, the mammalian neocortex is a large, multilayered part of the brain that can be subdivided into distinct "areas" that process different types of information.

In subsequent chapters, we will explore how the neocortex differs across different mammals. But it is worth pausing here to consider how such diversity comes about. Is it under genetic control, or is the cortex a blank slate that can be adapted to anything? And is the diversity endless, or are there tight constraints? One researcher who has studied the diversity of the neocortex is Leah Krubitzer. Over the years, her lab has studied a truly extraordinary range of mammals, from humans and cats to sheep and bats, and from opossums to hedgehogs and platypuses. They found that certain areas are present in all mammals. The part where information from the retina arrives—the primary visual cortex—is there in all species that have been studied, including the mostly blind naked mole rat. In animals that are blind from birth, the visual cortex still develops. Other parts of the cortex seem much more adapted to the particular ecology of their owners.

In one of my favorite examples, Krubitzer mapped the representation in the neocortex of the bill of the platypus. This animal explores the world through sensing electrical impulses generated by muscle contractions of other animals by using its bill. It turns out, this sensitivity is backed up by a large territory of the cortex devoted to processing information from the bill. In contrast, for the raccoon, to take a random example, it is information from the hand that gets assigned the most cortical territory.

The fact that blind animals still develop a visual area and the match between body and brain across animals might suggest that neocortex organization is purely genetically determined. It turns out, however, that experience is vitally important as well. Nowhere is this shown more clearly than in an experiment by Krubitzer's lab that she has called "lab rats gone wild." In this experiment, the descendants of laboratory rats were born and raised in a large field pen closely mimicking rats' natural habitat. After they matured, their brains were studied. Now, neuroscientists are pretty confident they understand a lot about the rat brain. Rats have been kept and studied in laboratories under various circumstances, including quite enriched environments, for many decades. Their neocortex has been carefully mapped, and we know exactly where which part of the body is represented. Not quite so, it turns out. When the wild lab rats' brains were studied, they looked very different from those of their lab cousins. For instance, the parts of the brain representing a wild rat's trunk, tail, and hindlimbs were much bigger. Turns out that rats in the wild climb and jump in a 3-D world, rather than the mostly 2-D lab environment, and this directly affected how much of the neocortex dealt with information from the relevant body parts, such as the body core used for climbing and balancing.

We could have known. Although an animal born blind develops a visual area, it is much smaller and poorly connected to the rest of the brain. We can see this type of change due to use in individual humans as well. Piano practice enlarges the representation of the fingers in the cortex, and fighter pilots trained to make quick decisions in response to lots of visual input show increased connections between parts of the cortex. Your genetics, your niche, your experience—everything influences the organization of your neocortex to some extent. It is, as Krubitzer termed it in one article, a magnificent compromise.

We know a lot about the neocortex of today's mammals, but do we know anything about what the neocortex of the first mammals looked like? This is where the comparative method shows its strength. By comparing the brains of many species and looking at elements the different brains have in common, we can put together a rough reconstruction of their common ancestor's brain. By applying this approach to the neocortex, we can infer the common ancestor of all mammals had a neocortex consisting of about 20 different areas, a bit like the current day shrew. For comparison, our human neocortex has about 180 areas. Our early mammalian ancestor probably had a few unimodal visual and somatosensory areas, but not yet an area specialized for motor control. It likely already had some associative areas as well. So quite a lot of processing power. The question that arises is what use these regions were for the mammalian common ancestor. To answer this question, we can take a closer look at what some of these regions do in our own, human brain.

Functions of the neocortex

In the early 1990s it became possible to study the working of the brain in living humans performing simple tasks. Prior to that, for most investigations about how the brain works we were reliant on studies in patients—war wounds especially provided localized lesions that told researchers a lot about what specific parts of the brain do—and studies directly interfering with the brain in nonhuman animals. Although we have been able to record the electrical activity of the human brain using electro-encephalography since the 1920s, this early technology did not have the resolution to tell us where in the brain the signals were coming from.

Positron emission tomography, or PET, changed this. The technique relies on the fact that when brains are active they require a lot of oxygen. This oxygen is transported via the blood to those parts of the brain in need. Thus, if you could find a way to measure the flow of oxygen-rich blood to each part of the brain under different conditions, say when people are looking at something compared to when they have their eyes closed, you would be able to infer which part of the brain is processing, in this case, the visual information. Measuring blood flow can be done by injecting people with a very low dose of oxygen-15, a radioactive isotope of oxygen that has a very fast decay rate. The oxygen travels in the blood to brain areas that need it. When the isotope decays, it loses positrons which can be detected by the PET scanner. We can then reconstruct where the most positrons came from, providing a 3-D picture of blood flow in the whole brain. If we now do this in two different conditions, say when people are looking at something and when they have their eyes closed, we can compare the two maps and see where there is more blood flow. In our example, those regions that have more blood flow during the looking condition are then inferred to be involved in processing visual information.

Having a technique to visualize the living, working brain revolutionized the field of neuroscience. Researchers quickly started to look for brain regions associated with specific functions. The field eventually became known as "brain mapping." Neuroscientists quickly started mapping out which regions of the brain were involved in basic tasks such as processing information from our senses, including the eyes, the ears, touch, and those controlling our muscles during movement. They could then compare what they saw in the living human brain with their earlier results from nonhuman animals or patients. Some researchers, however, chose to focus on more complex processes, such as thought, free will, decision making, and language. After all, these were the processes thought to be more developed in the human brain that could not be studied in other animals.

It has long been thought that what distinguishes humans from other animals is that they are less prone to simply respond to environmental stimuli, but rather have more control over their behavior. This is reminiscent of MacLean's "rational brain" idea. One of the first studies attempting to identify brain regions involved in these more high-level processes therefore aimed to study the neural basis of free will. In this experiment, participants climbed into the PET scanner and had to perform two types of tasks. In the first task people had to perform an action, such as lifting or flexing a finger, in response to a visual instruction. This was the simple condition, aimed at activating those parts of the brain involved in simple processes that any animal can perform. In the second task the participants had to perform the same movements, but had to do so randomly without any instruction. The idea was that the second task was one where people had to generate movements based on their own free will, rather than as a slave to instruction.

Emphasizing the grand topic of the study, the authors gave the paper reporting their results a particularly grandiose title: "Willed action and the prefrontal cortex in man: A study with PET," reflecting the terminology of the Victorian pioneers of science. As their title suggests, their result was that two regions in the most anterior parts of the human neocortex, the frontal cortex, showed more activation in the "willed action" condition. The first region is located near the outside of the brain in a part of the cortex that is specific to the brains of certain primates, and we will talk about it in chapter 4. The second region is located more toward the middle of the brain. This region is called the anterior cingulate cortex. Interestingly, this region involved in the grand behavior of free will is one of the regions that is present in all mammals and therefore likely was present in the mammalian ancestor.

Since the days of the early PET experiments, research studying the living, functioning brain has come a long way. Mostly, it now uses the technique of magnetic resonance imaging, or MRI for short. Rather than having to give volunteers a radioactive substance, MRI uses the different magnetic properties of oxygen-rich and oxygen-poor blood to visualize which areas of the brain are working hardest. Thousands of experiments are run this way each year. In quite a lot of them, the anterior cingulate cortex lights up. In fact, the anterior cingulate cortex—or ACC as it's known to researchers—is something of a Rorschach test for neuroscientists. They all see in it what they want to see.

ACC is activated in many experiments that require some form of conscious control. When volunteers receive information that they're making a mistake and adapt their behavior, ACC is activated. If they concentrate more on a task, ACC becomes more active. When they are performing a task that is more difficult, ACC becomes more active. If they even anticipate that an upcoming task is going to require more effort from them, ACC seems to become more active. As a result of detecting this region so often, researchers have postulated a wide variety of theories on what exactly this region might be doing. During my PhD research, I worked on theories that ACC was involved in detecting and helping correct and prevent erroneous behavior. I then moved to a lab that had proposed a role for ACC in determining how much effort somebody would be willing to spend on a task. More recently, my student and I have published a paper showing that ACC has a role in mediating social interactions with other people. To illustrate the variety of ideas on ACC, when I graduated from my PhD, a friend of mine gave me a framed spoof cover of a scientific journal advertising an article about "ACC as a homunculus," a little man in our brain controlling all our behavior.

Can our knowledge of mammalian evolution perhaps help us understand the role of ACC? Remember that early mammals opted for a very specific niche to escape the reptile dominance. Having small bodies and living at night afforded them living space but came at a cost. To keep up their body temperature they had to consume lots of energy. That is still the case. The

Komodo dragon eats about 80% of its body weight in a single meal, but the flip side of that is that it can survive on as little as one or two meals a month. This is not the case for a mammal like the shrew, which needs to feed regularly to keep up its energy demands.

These two different feeding habits have consequences for the foraging strategies. If the dragon doesn't get its prey today, that's nothing to worry too much about. There will be another day tomorrow. This is different for the shrew. It needs to be successful in obtaining food every day, or every night to be more accurate. Moreover, not every prey that wanders by might be good enough. If an easy to catch but not very nutritious prey wanders by, the shrew might need to forgo it for a harder to catch, but more nutritious prey. But the costs should not be so high that the shrew is likely to end up not catching any prey or using up more energy than it gains, or it might just not get enough nutrients in. This makes the foraging decisions of the shrew much more complex that those of the Komodo dragon.

ACC is one of the association regions that we find in all mammals and by extension we assume was present in the early mammalian ancestor. Could ACC have a role in making these foraging decisions? It certainly seems to be in a good position. It receives information from a lot of other parts of the brain, especially regions that convey information about knowledge of the environment not directly in front of us, such as the memory system, and information about value, from the reward system. In turn, ACC can provide information to the motor system, helping shape the movements the animal is going to perform.

When I arrived in Oxford in the lab of Matthew Rushworth for my postdoctoral research, I met Nils Kolling. Rather than starting his PhD as most students do by reading everything about neuroscience, he started by reading a standard text about foraging behavior. Through that he learned to understand how animals make decisions—what kind of information they use and how they compute their optimal behavior. Next, he devised an experiment to see if the ACC of human volunteers is able to represent the kind of information associated with the mammalian foraging choices.

Kolling asked volunteers to imagine they were monkeys foraging in a fruit tree. Initially, this is very simple. All you need to do is decide which of the many fruits available you want to consume. This is pretty much the type of binary choice that experimenters ask volunteers to make in countless psychological experiments every day, often to shed light on the complex role of ACC. After a while, the situation becomes more complex. By eating all the fruit from the tree, the tree becomes depleted. As a result, the tree becomes less attractive as a foraging patch. It might be time to look around for better trees that offer more and better fruit. Therefore, before the volunteers chose their piece of fruit, Kolling's experiment offered them a chance to move to another tree. They got a look at the potential other trees and were told how costly it would be to move to another tree. These are precisely the kinds of information a foraging mammal faces—do I stay here, or go to another foraging patch even if I'm not quite sure how good this is and how effortful or dangerous this is? So, on each trial, volunteers in the experiment first had to make a "foraging" decision (should I stay or should I go?) before they had to make a simple fruit picking decision.

Volunteers were performing this experiment while in the MRI scanner, so that Kolling was able to record the activation of brain regions throughout. Afterward, he looked for regions that showed signals related to the foraging decision. Through careful analysis, he could identify which brain region changed its activation based on the effort associated with moving to another tree, which region calculated the balance between the current tree and potential others, and that kept a record of the

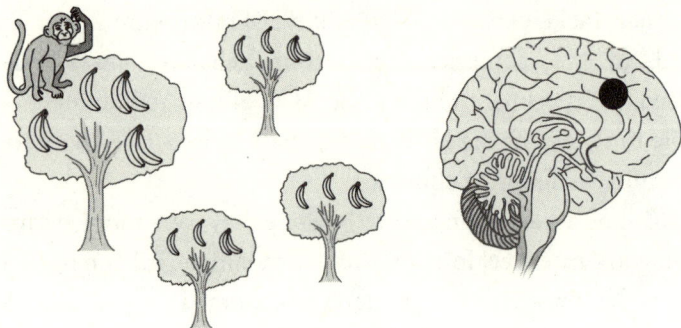

FIGURE 2.2. The experiment by Kolling. Participants imagined they were a monkey that had to forage for fruit in a tree. When the monkey had to decide whether to stay in the tree or find a better tree—a foraging problem—the anterior cingulate cortex (ACC), seen in the human on the right, is involved. But ACC is much less involved when the monkey is deciding which of the fruits in front of it to pick.

environment. Low and behold, it was ACC that seemed to form the hub where all this information came together. This ancient mammalian region performed some of the complex computations associated with the foraging challenges of early mammals. In contrast, the binary fruit picking decision was performed by completely different regions that evolved much more recently.

Kolling then decided to see if he could push this idea further by devising a new foraging challenge. One difference between the reptilian and mammalian foraging we outlined above is that mammals are under continuous time pressure. They have to find enough nutrients before time runs out, and they have to retreat to their safe harbors during the day. When time is running out, they might need to risk going after a more difficult prey or spend more effort to find a better tree, even if the chances of success are limited. It's literally all or nothing. So what if we put the volunteers in our MRI scanner under a similar pressure? If we tell them that they have to acquire a certain amount of "food" or points before a certain time or risk losing everything, will they adapt their behavior? This is exactly what Kolling did. Volunteers adapted their behavior in the way you would expect. Initially they were sensible and weighed the pros and cons of every decision, but when the deadline approached and they did not have enough food, they started taking more risks. Kolling calculated a complex variable to quantify how much risk would be beneficial. When he looked into the brain data to see if any region was representing such a variable, again ACC stood out.

Of course, decisions are not made just by one region of the brain. Contrary to my spoof scientific paper, ACC does not house an all-knowing homunculus that decides what we do. But it is intriguing that a region that was probably present in our common mammalian ancestor is able to integrate all the types of information involved in the specific challenges faced by that ancestor. Experiments similar to these two MRI studies have been performed in a range of mammals, including rats, mice, macaque monkeys, and humans. All of them use ACC to help decide about which action to take, depending on the current need and the effort and risks associated with the alternative. All this because of the cold, nocturnal niche of our mammalian ancestors.

––––––

Recap: The dragon and the shrew

Although the Komodo dragon looks like a monster from a 1950s horror movie and is still responsible for a very occasional human death, it is in a way a relic of a past time. Smaller reptiles live all over our planet in the present day and are perfectly well adapted to their niches, be they long-living turtles, venomous tropical snakes, or evolutionarily ancient but successful crocodiles. But the time of large reptiles dominating the landscape is gone. Ironically, it was fleeing to the nocturnal niche where reptile predators could not get to them that set off the mammalian journey to their current levels of dominance. The neocortex helped achieve that. It is not a completely new invention built on top a reptilian brain, but a modification that helps mammals reprocess information again and again. It also doesn't work in isolation, but together with all the other parts of the brain, some of which still perform their old functions as before, some of which got modified themselves.

After the extinction event that killed the dinosaurs, mammals diversified into many different lineages. All the major groups of mammals—monotremes, marsupials, and placentals—had already formed before the extinction event, when mammals were still hiding in the night, but once they were given free range, all the major branches of the mammalian family tree started sprouting new offshoots. The placentals form by far the largest group and are commonly subdivided into four superorders: the Afrotheria (aardvarks, elephants, manatees), the Xenarthra (anteaters, sloths, armadillos), the Laurasiatheria (bats, whales, hoofed animals, carnivores), and the Supraprimates (primates, rabbits, rodents). Each of these animals adapted their body and their brain in different ways, depending on the niches they occupied. We humans belong to a particular branch of placental mammals, the primates. As we will see, primates evolved to fill a very specific niche, one with a unique set of foraging challenges.

3

The squirrel and the squirrel monkey

THE PRIMATE VISUAL-MOTOR SYSTEM

Eastern grey squirrel

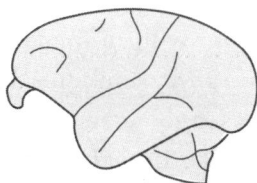

Common squirrel monkey

It is morning in the Colombian rainforest, the time when the temperature is still pleasant before the midday heat. A small monkey jumps out of the dense foliage of a nearby tree. It is very agile as it moves along the thin branches of the high trees. Its sharp eyes have spotted a tree bearing ripe fruit. It runs across the branches, then jumps across the large gap between two trees to land on the small branch below the ripe fruit. It clings to the branch using its feet and one hand, reaching out with the other hand to grasp the fruit. It takes the fruit between its hands, shifting its weight using its tail for balance, and starts peeling off the outer layer of the fruit to get to the juicy insides. Its moment of peace is suddenly ruined as the hoots of 10 to 15 of its conspecifics erupt. They have spotted the fruit as well. Their black and yellow faces give them a bit of the appearance of cute skulls, if there is such a thing. In Dutch the animal's name is therefore *doodshoofdaapje*, "death's head monkey." They are members of one of the most abundant primate species, the squirrel monkey.

Just as we explored the differences between mammal and reptile brains in the previous chapter, in this chapter, we will look at how the brains of primates began to differentiate themselves from other mammals. The squirrel monkey shares part of its name with an animal in my garden that superficially seems to live a very similar life. Like squirrel monkeys, this animal also jumps easily across the small branches of trees and lives off their products. The animal in my back garden, however, is not a primate, but a rodent.

The gray squirrel is an extremely successful, mostly arboreal, animal, indigenous to North America. Since its introduction into Britain, it has largely replaced the indigenous red squirrel and is now considered a pest by many. Cute rats, as one colleague described them to me. The squirrel is not alone in being so prevalent. Rodents have become one of the most successful orders of mammals, representing about 40% of all living mammals. They are generalists that easily take over a very diverse range of niches. Their diverse membership includes the giant capybara that inhabits savannas and dense forests in South America; the tiny kangaroo rat that prefers the semiarid deserts of North America; the semiaquatic beaver; the common mouse, appearing pretty much anywhere where humans live; and the gray squirrel terrorizing my garden trees.

The comparison of the squirrel monkey with its rodent namesake seems accurate. At first glance, the two animals are very alike. But primates and rodents took very different evolutionary trajectories, which are evident when we compare the namesakes a bit closer. Primates' bodies have particular adaptations both to sense and to manipulate the world. When it comes to sensing the world, the eyes of the primate are different from those of many mammals. Primates tend to have relatively large eyes that are placed to face forward with a large overlap in visual fields. This increases the binocular vision which helps the animals see depth. It also helps the animal to perceive more of an object even if it is partly obscured behind an obstacle, and it enhances the ability to detect faint objects. By contrast, the squirrel's eyes are placed more on the sides of the head. This gives the animal a wider range of vision, which is helpful to detect predators. The primates' large, forward-facing eyes are more useful for moving through or manipulating the environment than the squirrel's eyes. Nocturnal primates often have very large eyes and in some cases a light-reflecting layer on the retina, as with cats. Many of the primates that are active in the daytime have a specialized area in the retina where light-sensitive cells are packed very closely together, giving good visual acuity, and they have good color vision. Like all mammals, primates also rely on hearing and smell, but their visual adaptations are unique.

The way primates move is also different from rodents. There is great variability among living primates in how they move through the environment, including leaping between trees; climbing along continuous branches using all four limbs; suspension from branches and swinging between them; walking on the ground using all four limbs, including walking on the knuckles of the front hand; and the human approach of walking on two legs. These different forms of locomotion are all associated with specific adaptations of the skeleton. Nevertheless, some commonalities distinctive of primates are apparent. Primate skeletons are relatively primitive, retaining many characteristics of early mammals, but primate hands are quite distinctive. Primates have five digits with nails rather than claws. The four fingers of each hand can be flexed and extended and moved a bit from side to side, but it is the thumb that literally stands out. The joint at the base of thumb, connecting the metacarpal bone to the rest of the hand, allows the more complex movements associated with grasping things. Although the squirrel is capable of holding a nut between its two hands and bringing it up to its mouth, this is nothing compared with the ability of the squirrel monkey to hold the fruit in one hand or to manipulate it in preparation of consumption. You will never see a rodent peel the skin of a banana.

We do not know the exact lifestyle of early primates that gave rise to these adaptations. True primates first appear in the fossil record about 50 million years ago, but molecular evidence suggests that primates first appeared 85 million years ago, before the extinction of the dinosaurs. That leaves 35 million years of unexplained primate life. But we have some clues. Given the niches of most mammalian species at the time, primates likely evolved from a small nocturnal mammal, possibly an insectivore. Their skeleton seems adapted to life in trees. Early hypotheses on primate evolution suggested that early primates were particularly adapted to an acrobatic way of moving by using their legs to push off from tree branches to leap to another tree and using their hands to then grasp something to hold on to. Such a lifestyle accounts for their grasping hands without claws and some of the specializations in the feet and knees. Other theories point out that primates have many features not seen in other arboreal animals, such as the forward-pointing eyes for stereoscopic vision. This is characteristic of visual predators, such as owls and cats. Maybe early primates were predators stalking prey in the canopy of forest trees. Others have argued that early primates were most likely to be omnivores, rather than predatory carnivores, because all current insectivores among the primates are extreme specialists. An alternative scenario is that primates evolved in conjunction with flowering plants and exploited their products in a small branch setting. Whichever proves to have been the correct order of events, all theories agree that primates are animals that originally evolved a behavioral repertoire centered around complex visually guided movements. In other words, these are animals for which vision is the most important sense and who have developed specific movement abilities, particularly in the domain of grasping.

The heritage of these adaptations can be seen in all present-day primates, no matter how different they are. Primates nowadays are found in a variety of habitats, ranging from African deserts to the cold mountains of Japan, but they tend to mostly prefer tropical forests of one sort or another. They vary in terms of the forest type they like, ranging from dark primary forests, simpler secondary forests, woodlands of trees and bushes, or even the more open savanna. Within a particular forest, different species also prefer different locations, from the dark understory, via the main canopy, to the tops of the trees where light is abundant. This wide variety of habitats leads to an equally wide variety of lifestyles. Primates differ widely in the range of territory they visit, activity pattern over the day and night, diet, and social structure. But their heritage is present in all and reflected in all their brains. In this chapter, we will start investigating how the primate brain became specifically adapted to sight and movement through the small branches of trees—a visuomotor brain.

———

Two visual pathways in the primate brain

The brains of the squirrel monkey and the squirrel in the figure at the beginning of the chapter look quite different. Both have the "cover" of the mammalian neocortex that hides most of the rest of the brain from view, but in the primate this cover is much more extensive and folded. Folding is nature's way to pack more surface area into a confined space; the squirrel monkey's neocortex is thus even bigger than it looks. Bigger bodies often coincide with bigger brains, so perhaps this is an unfair comparison, but even smaller species of primate have larger brains than similarly-sized rodents. Armed with impressive mustache-like facial hair, the tamarin, a small monkey from Central and South America, is about the size of a squirrel. The tamarin brain is not quite as folded as that of the squirrel monkey, but at 10 grams it still beats the squirrel's 6 gram brain. Compared with most other mammals, the primate brain is bigger for a given body size.

But is weight the best measurement of brain power? It would be even better if we knew how many neurons each of these brains possesses. Counting all the millions of neurons in a brain one by one might take a while, so we would have to come up with something clever. This is precisely what the neuroscientist Suzana Herculano-Houzel thought. She had been puzzled by the fact that neuroscientists often quote the number of neurons in the human brain as about 100 billion, but that there actually seemed to be very little data to prove that number. The most sophisticated methods rely on slicing brains into thin sections and counting the numbers of cells in small sections of these slices. Then the total number is estimated by simple extrapolation.

The problem with this method is that neurons are highly unevenly distributed in different parts of the brain. Instead, Herculano-Houzel came up with a different idea: to turn the brain into soup. If a researcher wants to preserve a brain after the owner has died, the brains are often fixed using paraformaldehyde. This has the effect of strengthening the cell nuclei. Then, using a chemical detergent, Herculano-Houzel dissolved the rest of the cells, such as the membranes of the cell itself and the cell bodies such as mitochondria. Only the cell nucleus remained. Now, all she had to do was extract the undissolved cell nuclei and count them in a known fraction of the solution. Using another chemical agent she could label which of the nuclei belonged to neurons and which belonged to other cells that support the neurons, such as glia. Voilà, she had a method to count neurons. Now she could explore any brain, from any species, and any specific part of the brain. One of the first stops was to compare rodent and primate brains.

Primate brains turned out to be even more impressive than previously thought. Not only are the brains bigger for a given body size, primate brains pack many more neurons per unit of space than rodent brains. The squirrel has about 77 million neurons in its cerebral cortex, the tamarin probably between 250 and 350 million. Moreover, neurons in the primate neocortex are efficiently packed in, so much so that a 10-fold increase in neurons will need a 10-fold increase in neocortex size in primates, but a massive 50-fold increase in neocortex size in rodents. Compared with rodents, primates can pack a lot more computational power into their brains. But computational power in itself doesn't say much about what the brain is specialized for. It doesn't relate in any particular way to a specialization for visual and movement skills. For that, we need to look at the internal organization of the brain.

The brain of primates has a lot of territory devoted to processing visual information. As we discussed in chapter 1, even many evolutionarily old parts of the brain receive visual information, for instance to keep track of the cycle of night and day. In the primate neocortex, according to some estimates, up to 50% of brain areas in one way or another process information from the eyes. Visual information enters the neocortex at the back, in the occipital cortex. Two-dimensional images from the retina are carried to the primary visual cortex.

One of the major tasks of the neocortex is to convert this two-dimensional representation into various three-dimensional ones. Although we perceive our visual environment as perfectly coherent, different aspects of the visual world get processed in very distinct parts of the cortex. From the occipital cortex, two major pathways carry visual information to the rest of the cortex for further processing. One pathway runs from the occipital cortex to the temporal cortex, the part of the brain just above our ear. The other pathway carries the information upward, toward a part of the cortex called the parietal cortex. These temporal and parietal pathways work in very different ways.

In the 1970s and 1980s, researchers at the National Institute of Health in the United States investigated the processing of visual information in the cortex of the macaque monkey. Macaque monkeys are so-called Old World monkeys, which means that they naturally live in Africa and Asia (and, in the case of the barbary macaque, on the small tip of Europe on the mountain of Gibraltar). They share a common ancestor with humans from about 29 million years ago. They are clever animals that can be taught to perform complex tasks in a laboratory setting.

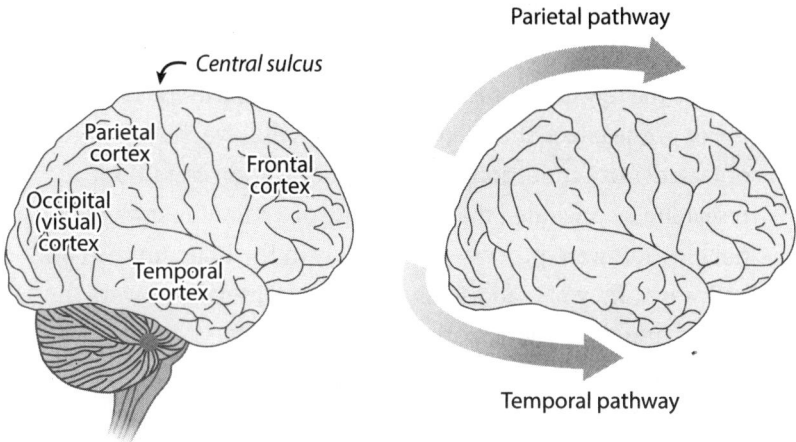

FIGURE 3.1. Two pathways transport visual information from the primary visual cortex at the back of the brain to the temporal cortex behind the ear and the parietal cortex at the top. From the parietal cortex, information goes to the frontal cortex of the brain, including the motor areas that are able to control the muscles.

In this case, the researchers taught them to choose between different objects. In repeated tests or trials, they had to choose one of two objects and were either rewarded or not. The monkeys were able to learn which of the objects was the rewarded one and would choose it reliably. In a variant of the experiment, the identity of the object was not important, but it was the location that was rewarded. In this situation the monkeys quickly learned to choose the object located close to a visual landmark, independent of its identity.

To study how visual information is processed in the macaque brain, the researchers gave some monkeys selective lesions of either parts of the temporal cortex or the parietal cortex. The results were striking. Monkeys with temporal lesions lost the ability to discriminate between the different objects. They were still able to choose successfully in the location condition, so they were not blind and they had not lost the ability to make decisions. But they were not able to distinguish the objects and choose the rewarded one. Monkeys with parietal lesions showed the opposite effect, they could perfectly well solve the task in the object identity condition, but not in the location condition. The researchers concluded that the quality of a stimulus and its spatial location were processed separately in the brain, in the two pathways. They named the temporal and parietal pathways the "what" and the "where" pathways, respectively.

The functional dissociation between the two pathways was striking, but to some researchers also unsatisfactory. Perception alone provides no evolutionary benefit. Processing the "what" and "where" of visual information is useful only if this information is used for something, most likely to achieve some sort of change in the environment that will have a benefit to the organism. In the previous experiments, the type of response the animal had to make, reaching for the right object, was always the same. Indeed, many experiments on human perception of the environment do just that, keep the responses as similar as possible and vary only the visual input. The assumption is that any type of visual input can be transferred into movements in the same way. However, evolutionarily speaking, it makes much more sense to look at the entire pathway—from input to output—as the target of selection, since only a complete system can be formed, not just a "half" perceptual system that then hooks up to a generic output.

Looking back, this kind of dissociation between visual systems is present in all vertebrates. Different pathways processing different aspects of visual information are apparent from the moment light hits the retina. Information from different types of photoreceptors is combined by different populations of neurons. By the time information is transmitted via the optic nerve from the eye to the brain, different signals convey the spatial locations of light hitting the retina and the temporal profile of stimuli, among others. Information from the eye does not just go to the cortex but is diffused to a range of target areas. Many of these systems likely evolved independently, each to solve different problems faced by their owners. This argument was made by the neuropsychologists David Milner and Mel Goodale in their book *The Visual Brain in Action*.

As an example, Milner and Goodale referred to research done in the frog. Normally, these animals pursue prey objects, such as mealworms, when the prey is in the frog's visual field. After a lesion of the tectum, a part of the midbrain, they show no interest in them at all. The same happened when a moving disk was moved quickly in the direction of their eye, simulating a predator moving toward the frog. Healthy animals jumped away while those with a lesioned tectum did not respond. The frogs could still respond to predators, though, because the lesioned animals did jump away when an experimenter touched them on the limb. But their jumps conveyed something interesting.

If there was a barrier near the frog, it would carefully avoid jumping into it when trying to avoid the predator. When a frog with its eyes surgically closed was touched on the limb, it jumped forward, even if there was a barrier in front of it. Without eyes, it could not see the barrier. An animal with functioning eyes but with a lesion of the tectum, which seemed unable to visually register predators, nevertheless avoided the barrier when unexpectedly touched on the limb. The frog did not visually register the predator, but it did visually register the barrier. Thus, seeing seems to mean different things at different times. Different parts of the brain use visual information to detect predators and to avoid obstacles. When the predator detection pathway is damaged, the animal can still avoid obstacles.

Milner and Goodale argued that the crucial difference in visual processing seen in the frog experiment is not just the type of visual stimulus, but also *how* the information is used. Could this be the case for the temporal and parietal pathways seen in the primate, including human, brain as well? They studied patients with lesions of the temporal and of the parietal cortex. One such patient was a woman identified by initials D.F. She had suffered carbon monoxide poisoning that left her with extensive damage to her temporal cortex. When asked to report on the shape of simple objects placed in front of her, she reported to be unable to describe them. Shown a picture of an apple or a book, she could not recognize the object and could not draw a copy of the picture. However, when asked to draw an apple or a book from memory, she was able to do so, indicating that she had not lost the understanding of what the objects were or the ability to perform the actions to draw them. But shown her own pictures later on, again she did not recognize them. The ability to consciously recognize objects was lost in D.F.

But D.F. had not lost all uses of visual information. When given a card and asked to put this into a letterbox slot, D.F. rotated the card perfectly well to match the orientation of the slot. Even if the letterbox was rotated, she could always post the card. She started rotating the card into the correct orientation as soon as she started the movement, as healthy people do. However, when asked to describe the orientation of the slot, again D.F. had no idea. A similar thing happened in a task in which she was given rectangular objects and asked to pick them up with the thumb and middle finger. She anticipated the width of the objects with her fingers, just as healthy people do, but she was not able to indicate the width of the object with her fingers to the experimenter—a task for which she had to make a perceptual judgment about the object.

D.F.'s behavior, caused by damage to her temporal cortex, contrasted with that of another patient, V.K., who had damage to her parietal cortex. Whereas healthy people shape their hand and fingers to suit the object they want to grasp over the course of their reach for the object, V.K.'s grasping showed she failed to match the distance between her fingers and the object appropriately. Other patients with parietal damage showed a variety of disturbances in either reaching toward objects or shaping their hand to grasp it appropriately.

Based on these types of results, Milner and Goodale concluded that the temporal cortex might be essential for perception of the outside world, for instance, when we have to make judgments about aspects such as color, form, or texture of an object, while the parietal cortex is important for processing visual information needed to guide one's motoric actions. In other words, the parietal pathway is the most important for the type of behavior we identify as typical of primates: visually guided actions.

The parietal pathway set primates up for a very particular way of viewing the world, one that still affects us humans today. We will explore this pathway in the remainder of this chapter and come back to the temporal pathway in the next.

———

Vision for movement

What exactly does the parietal pathway do? To effect change in its environment, an animal needs to perform an action. For this to happen, the visual information that hits the retina must in some way be transformed into contraction and flexion of muscles. To see how complex this transformation is, let's look at our squirrel monkey again. Suppose our squirrel monkey wants to grasp a nice, juicy fruit located on a branch just to the left and above him. He can view this object in a visual, eye-centered reference frame; however, he might not be fixating on the object, for instance if he wants to keep an eye on a fellow monkey that might also want the same fruit. He has to calculate a transformation from the position of his hand to the fruit, while the center of his visual fixation is the other monkey. This is complicated by the fact that he is standing on a flimsy branch that moves in the wind, continually changing his visual reference frame.

When the other monkey moves and our squirrel monkey moves his eyes or head to keep it in focus, again his visual field changes, although the fruit is in the same place in the real world. Still, he has to convert the planned movement of his hand to the fruit into muscle coordinates, for which he has to compute which muscles to contract and which to flex. While he does this, he feels an insect land on his elbow and moves his arm to scare it away. Because his arm is now at a different angle to his body, the movement toward the fruit now involves a very different set of muscle contractions, even though his hand and the fruit are still in the same place. To accomplish a successful movement, somehow the brain needs to align all these different coordinate systems—retinal coordinates, head position, body position, relation of the limbs to the body, and state of muscles—to transform information from one system to another. One of the main functions of the parietal stream is to perform the necessary computations.

In many primates, the neocortex shows a big groove in the middle running from the top downward and dividing it into a posterior (back) and anterior (front) part. This kind of groove in the brain is called a sulcus. Because of its location, the groove dividing the neocortex is termed the central sulcus. The area behind the central sulcus is the parietal cortex. The parietal cortex receives sensory information from the body directly from the spinal cord. For instance, when an area of the body is touched, part of the anterior parietal cortex will register this. The parietal cortex receives visual information from the occipital cortex; this is part of the parietal pathway we talked about earlier.

As a whole, this means the parietal cortex processes information about the state of our body, including how the body is oriented, which muscles are currently active, and what the effect of flexing or contrasting muscles would be, and it processes information about the visual state of the world, with the information coming from the visual cortex, including information such as the location, size, and shape of objects to interact with. Together, these pieces of information allow the parietal cortex to estimate the current state of all relevant aspects of the body and the world that need to be integrated for us to successfully act upon that world.

Having access to the position of the body and the visual information is one part of a successful visual-to-action transformation. The other is to affect the muscles themselves. This is largely the role of the territory directly in front of the central sulcus, called the frontal cortex. The posterior part of the frontal cortex has direct connections to the spinal cord. These connections reach neurons in the spine, some of which send connections directly to the muscles of the body. This means that through their connections with the spinal cord, the areas directly in front of the central sulcus can influence the effectors, that is, the eyes, body, limbs, and fingers that affect the actual change in the world. Hence, this part of the brain is called the motor cortex. The motor cortex gets information about the state of the world through direct connections with the parietal cortex. Together, parietal and motor cortexes act as a "visual-motor" stream, processing visual information, converting information across different reference frames, and ultimately influencing the muscles.

Although present in other mammals, including the squirrel, this visual-motor stream consisting of parietal and motor cortex is vastly expanded in primates compared with other mammals. The part of the parietal cortex that receives visual information has expanded so much over evolutionary time that it makes up the vast majority of the parietal cortex in primates. The primate motor cortex has similarly expanded. In fact, certain parts of the motor cortex might be present only in primates. One such region is located quite low down in the frontal cortex and is important in movements of the head. This fits well with the idea that primates are the only arboreal animals that make particularly complex movements of the head when trying to reach for the products of the fine branches of trees. The visuomotor pathway might not be a truly unique primate adaptation, but it is certainly much larger and much more complex in primates than in, say, the squirrel.

One way to characterize the specific "primate-ness" of the squirrel monkey's motoric abilities is their variability. Jumping, climbing, grasping, manipulating, and balancing together make a much more varied repertoire than that of the squirrel. Can the organization of this large visuomotor expanse in the primate brain tell us anything about how this variety is achieved? Parietal and motor cortex are both not indivisible wholes. In fact, lesions to different parts of posterior parietal cortex can selectively interfere with almost every aspect of visuomotor behavior, including eye movement, posture, reaching movement, hand shaping, and force control. The same case can be made for the motor cortex. Cells in different parts of the motor cortex seem to fire during different types of movements. The cells firing when the monkey makes head movements in the lower part of the motor cortex are just one example. In fact, the visuomotor pathway seems to consist of a number of parallel subpathways connecting specific parts of parietal cortex to specific parts of motor cortex. Whatever principle of organization describes how these parallel pathways differ will tell us a great deal about how the primate achieves its distinctive abilities.

———

Grasping the parietal pathway

Generations of neuroscientists have grown up learning about the "motor homunculus." This is the idea that the motor cortex, the area just in front of the central sulcus, contains a "map" of the human body. Each part of the human body is controlled by a different part of the motor cortex. If a body part requires particularly fine control, such as the primate hand and fingers, it will be assigned more territory. The areas controlling different parts of the body together form a kind of map on the cortex. This map is often represented as a homunculus. A similar map was proposed to lie just behind the central sulcus, representing the areas in which information from the body was received.

The motor homunculus was identified by electrical stimulation of the cortex. Stimulating the brain and recording which muscles contract is an easy and relatively cheap way of exploring the brain. In the 1870s, the German scientists Fritsch and Hitzig pioneered the use of this technique on the dog brain. Experimenting at home, they opened up the skulls of dogs, exposing their cortex. They then stimulated different parts using

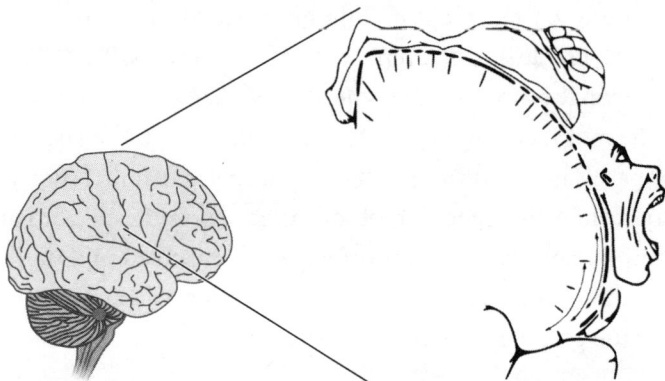

FIGURE 3.2. The Penfield motor homunculus. Stimulating different parts of the motor cortex leads to twitching of muscles in distinct parts of the body, represented here by the body parts. The size of the displayed body part represents the amount of brain tissue devoted to it.

a brief electrical discharge. They found that different parts of the frontal part of the cortex elicited twitches in the neck, foreleg, hindleg, or face of the animal. Following these pioneering studies, other researchers extended the technique to study other animals including, eventually, macaque monkeys using a variety of stimulation techniques. Instead of the short discharges used by Fritsch and Hitzig, others used longer stimulation durations. They saw much more complex movements than simple twitches. Over the decades different bodies of work emerged, with researchers debating the right way to study the motor cortex.

The debate centered around two distinct principles of organization of the excitable part of the brain. One group of researchers argued that the motor cortex is organized according to what they called effectors: each separate part of the brain controls a specific muscle. Other researchers argued the brain is organized according to motor programs, groups of muscles working together to form a movement. The first type of work eventually led the neurosurgeon Wilder Penfield to publish the motor homunculus, based on stimulation of the exposed cortex of human patients undergoing brain surgery. It provided a nice, clear map of the motor cortex that could be digested by generations of students.

Penfield's map was unsatisfactory from the start. It is a simplification. Penfield himself did not mean for his map to suggest that there was no overlap between the representations of different muscles on the cortex. Over time the nuance was forgotten. Researchers did, however, continue to report on experiments not consistent with the pure maps, questioning the clear separation between effectors and also noting that sometimes an effector could be represented in two different places.

One particularly insightful study came out in 2006, when I was performing my own stimulation experiments as part of my postdoctoral training. This study showed the complexity of the motor representations. The researchers injected the muscles of a monkey with a so-called tracer, a virus that moves backward through the axons toward the neurons in the spinal cord and then toward the neurons in the brain. By observing which neurons are infected by the virus, you can create a map of the connections of the injection site. The researchers observed that tracers from a single muscle ended up in a widespread part of the primary motor cortex, not in a single cluster. Moreover, tracers from different muscles ended up in overlapping populations of neurons. This showed that the clear separation of effectors implied by the simplified interpretation of the motor homunculus is wrong.

A few years before this 2006 study, the neuroscientist Michael Graziano and his colleagues had reignited the debate on motor cortex organization. In one experiment, one of Graziano's colleagues was attempting to find the region in the cortex that controls eye movements. Rather than stimulating the motor cortex with a brief 50 millisecond pulse, he used a high-frequency burst of about half a second, which he thought more closely resembled the natural duration of an eye movement. In one experiment, he missed the eye movement area and ended up stimulating an area of the motor cortex. What he saw astonished him. Rather than a twitch in a selective muscle, the monkey made a completely natural movement. It moved its arm in what looked like a perfect reaching movement. This was not a flexion of a single muscle, but a coordinated movement of the muscles of the arm. When Graziano and his colleagues explored further and stimulated another location, the monkey moved its hand to its mouth in what looked like an eating movement. If the arm was moved to a different starting position, stimulation of the same location still brought the hand to the mouth—a different coordination of muscles, but the same functional result. Moving their stimulation across the motor cortex, Graziano and his collaborators found a whole repertoire of movements.

The movements they found were all part of the natural repertoire of the monkey. For instance, stimulation of the area just in front of the central sulcus, about halfway between the highest and lowest point of the sulcus, caused the monkey to move as though manipulating an object in front of itself, at the height of the chest. Most movements involved the arm and hand, which would be expected given the movement repertoire of the monkey, but some involved a wider combination of effectors. Stimulation of the motor cortex toward the middle of the brain elicited arm and leg movements resembling leaping or climbing; stimulation of an area low down on the outside surface evoked defensive movements, including blinking, squinting, moving the head away, and lifting the hand as if to protect the face.

It seemed that an effector-specific organization of the motor cortex is only part of the story. The cortex seemed organized in terms of functional categories of movements. Rather than each region controlling one muscle only, each region controlled all the muscles important for a particular type of movement. All of these were movements that are part of the natural repertoire of the monkey. While there was a region that seemed to control object manipulation in front of the monkey's body, there was no region that elicited a movement like that in front of the face, because monkeys generally do not hold an object in front of the face when manipulating it.

As we discussed above, the motor cortex forms a series of pathways with the parietal cortex. The next step was therefore to stimulate regions in the parietal cortex to see if they could obtain similar results. The defensive area of the motor cortex gets its information from a specific part of the parietal cortex. This region was known to respond to visual stimuli near to or moving toward the face, but also to touches of the same area. It also seemed to respond to vestibular information, for instance when a person ducks or otherwise moves suddenly. One hypothesis was that this is related to defensive movements. Lo and behold, stimulation of this area elicited facial twitches, contraction of eyelids, and hand movements protecting the face, just as was seen after motor stimulation.

Follow-up work by other groups has since shown that this pattern holds. Stimulation of both parietal and motor regions can elicit natural movements. Researchers also found that regions in the motor cortex and in the parietal cortex that elicit the same movement type tend to be connected. The mystery of the parallel pathways was therefore solved. The parietal-motor system consists of a number of parallel pathways that are each involved in transforming visual information into muscle coordinates for a particular category of movement. The primate brain contains a highly specialized series of visuomotor transformation circuits, all adapted for life in the trees.

———

From a visual pathway to intelligence

The primate brain is a visual brain. As we have seen, though, visual information is not processed as in a single, coherent fashion in the brain. Parallel pathways evolved that each analyze different aspects of visual information and use that information to achieve a certain goal in the world. This is true in old evolutionary pathways that still help the present-day frog jump away from predators and not jump into obstacles. It is also true within the primate neocortex, where the temporal pathway processes how an object looks, but the parietal pathway processes where it is and how we should shape our hands and limbs to interact with it. Within the parietal pathway, specific parts of the parietal cortex interact with different parts of the motor cortex to achieve different types of movements, all of which helped our early primate ancestor navigate life in the small branches of trees: leaping from branch to branch, defending the body from harm, reaching and grasping for fruit, and peeling the fruit to get to the juicy insides.

As we found, the parietal pathway is largely a primate specialization. Something reminiscent of it is found in other mammals, such as our squirrel, but it has massively expanded in early primates. And it did not stop there. These visuomotor adaptations are so dominant in our brain they formed the basis of much of our other behaviors. Within different parts of the primate family tree, the parietal-frontal system expanded dramatically. If we compare primate brains of different sizes, we see that not all parts of the neocortex expand equally, there are "hot spots" of expansion. One of these hot spots is in the parietal cortex.

This is also true for the human brain. As the parietal pathway expanded, it became able to support other cognitive functions. If we look at activity of the brain when humans are performing different types of cognitive tasks, we see that this expanded parietal cortex is not just involved in visual-to-motor transformation. Rather, it seems to be activated by all kinds of cognitive tasks. Apparently, whatever kinds of computations this expanded parietal cortex can perform, they seem useful in a wide range of abilities.

When some part of the brain is involved in a lot of functions, it is often tempting to say it must have something to do with intelligence. Intelligence is an elusive concept, which can mean many different things to many different people. In fact, it has long been debated whether there is even such a thing as a single intelligence, or many different types of intelligence. These days the popular science and self-help sections of our bookstores contain a plethora of books on artistic intelligence, emotional intelligence, musical intelligence, bodily-kinesthetic intelligence, social intelligence, naturalist intelligence, and so on and so forth.

As in most good psychological discussions, the answer to the question whether we have one or many types of intelligence lies somewhere in the middle. Yes, different people can have different types of abilities moderated by distinct parts of the brain, but there is also a consistent factor that seems to explain some level of performance across all these different abilities. This consistent factor was discovered in the early twentieth century by the psychologist Spearman and termed the g factor. Many

FIGURE 3.3. The Raven's progressive matrices test. Participants need to determine the shape of the missing figure in the bottom left, based on the horizontal and vertical patterns.

tests have been derived to measure this factor, but the most consistent measurements involve quite dynamic tasks, in which people have to solve cognitive puzzles by structuring incoming information into understandable patterns.

The cognitive neuroscientist John Duncan at the University of Cambridge has long been fascinated by the brain mechanisms behind the general intelligence factor *g*. He has investigated the activity of the brain while participants solve puzzles that are meant to tap into this general intelligence. One such puzzle is Raven's Progressive Matrices test. In this test, the participants are presented with a series of figures and have to find the shape that will fit in a missing square. To do so successfully, they must integrate various sources of information, including size, shape, and color of multiple surrounding squares.

When participants perform this task, their brain shows increased activation in a network of regions, including the parietal cortex and parts of the frontal cortex. In fact, these regions seem to be activated by a variety of cognitive tasks that all require non-routine—"intelligent"—behavior. Duncan termed this network of brain regions the "multiple demand network." The multiple demand network seems to be involved every time the task requires some aspect of g. Something about the computations of the parietal cortex—a system optimized for online calculations of the quantities required for transformations of visual to motor information, such as location, shape, distance, duration—seems useful for solving fluid intelligence tasks such as the Raven matrices test. This sets up a theme we will encounter again and again in evolution of the neocortex, that the brain handily hijacks existing structures and abilities for novel uses.

For the quantities calculated in the parietal cortex to be useful in solving the Raven matrices test, it is not sufficient to relay the information only to the motor cortex. The calculation of which shape is missing relies on some of the same types of quantities as a motor transformation, but it is not a direct visual-to-motor transformation. Rather, it is a more abstract transformation that later has to be transformed into a motor response, usually the selection of one or multiple alternatives, that is quite separate from the "intelligence test" part of the task. The expanded parts of the parietal cortex therefore do not relay this information directly to the motor cortex, but to more anterior parts of the frontal cortex that deal with information of increasing abstraction and complexity.

This part of the frontal cortex, often termed the "prefrontal cortex," also expanded dramatically in some primate lineages. It did so just as parietal cortex and, it turns out, just as parts of the temporal cortex, the other visual cortical pathway. Together, these expanded parts of cortex allowed a distinct group of primates to forage, think, and behave in a whole new way. That is the next chapter's story of the simian primates.

———

Recap: The squirrel and the squirrel monkey

Although primates are not the only mammals living in the trees, their early representatives were uniquely adapted to the niche of living along the small branches. They developed a body and brain specialized for leaping from branch to branch and grasping the leaves and other products of the tree. The neocortex of the primate brain contains two large pathways processing visual information. One of those, through the parietal cortex, helps guide the unique motor repertoire of the primate. In some primates, this pathway expanded and was recruited for other behaviors. Some of the computations required for primate motor behaviors such as grasping turned out to be very useful for other so-called intelligent behavior. This shows how evolution opportunistically adapts what is already there.

Primates diversified early in their evolution into distinct lineages. Although all primates share the two visual pathways, there are differences in how they evolved in the different groups. In one group in particular, parts of the cortex underwent repeated expansions and reorganizations. The second visual pathway, through the temporal lobe, and the frontal cortex both expanded when a particular group of primates had to deal with some major changes in the Earth's climate. The warm environment in which early primates thrived was about to end.

4

The lemur and the macaque

THE SIMIAN WAY OF FORAGING

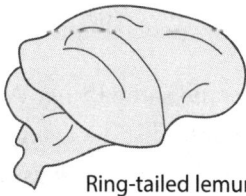

Ring-tailed lemur   Rhesus macaque

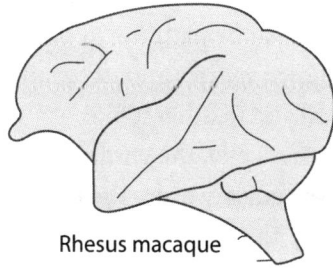

Not all primates are the same. From primate origins as arboreal visuomotor experts, some invested in even bigger brains, with the temporal and frontal parts of the neocortex expanding to support types of behavior that form the basis of some of our most characteristic human behaviors.

I am a fan of Formula 1 racing cars. I'm one of those enthusiasts who can look at a racing car and quickly tell you the era when it raced, what its design philosophy was, and probably more details about its aerodynamic performance than is healthy for the conversation. Of course, this is learned expertise. I wasn't born with knowledge of Formula 1 cars but acquired it over countless hours of watching F1 on television and peering through books and magazines. Needless to say, some people around me can get a bit exasperated about this. But other people have similar skill sets. I know people whose interest is not in F1, but gardening. Wherever they are on a walk, they can identify the type of any plant just by looking at it. This is not because they have simply learned to recognize every single plant in the area, but because they can classify plants using a range of factors, regardless whether they have seen it previously. It is astonishing the little details they can pick up from the shape of a leaf or patterns in bark that allow them to group an unknown species into its family group and then infer what kind of soil and weather it likes.

Basically, the gardeners and I both do the same thing. We very quickly order the information we get into distinct categories. These categories share certain properties that allow us to classify data. Through this mechanism, we can know a lot about one specific object when we can file it in its appropriate category. This ability to categorize information is one of the bases of human expertise. You have probably heard about the farmer who can recognize all his individual cows, even though they all look pretty much the same to anybody else. When asked how he does it, he probably won't be able to tell you. He just does after working with cows for all these years. When I was a PhD student, we had a technician in the lab who built the various devices we used to collect reaction times from volunteers in our experiments. All these devices looked the same to me, but this technician was able to instantly see "that was the one I built in April last year." Maybe he was bluffing, but I never caught him out. In the same way, you can likely recognize hundreds of

different faces without having to explicitly remember idiosyn-
cratic features of each of them. You are able to put the whole
together and store it in some way. As we will see, this ability to
quickly categorize information was very useful to some of our
primate ancestors and forms the basis of some of our human
abilities.

Another advantage we share with our primate ancestor is the
ability to learn about learning itself. This is best illustrated by a
nice experiment originally devised by the psychologist Harry
Harlow. He gave macaque monkeys a simple task. They were
presented with two objects. If they chose one they got a reward,
if they chose the other one they would get nothing. Then, they
were presented with the same two objects again. The same object led to a reward, the other again did not. And so on. Eventually the monkey figured out which of the two objects led to the
reward and would keep choosing that object. Then, after a few
more tries with the same two objects, the monkey was presented with two new objects. Again, only one was rewarded and
again the same two objects were presented a couple of times.
Now, if you understand how this task works, you will always
choose the correct object from the second time you see a pair
and so on. After all, if you were rewarded the first time, you keep
choosing that object. If you were wrong the first time, you
choose the other object and stick with that.

The question in Harlow's experiment was not whether the
monkey would eventually choose the correct object, but how
quickly it would figure out the rule and apply this perfect win-
stay/lose-shift strategy. Macaques can figure out this strategy.
After about 400 sets of objects they chose the correct object on
the second try almost 90% of the time. Not perfect, but some
evidence that they figured out the overarching rule guiding the
experiment. Humans can figure this out almost instantly. Just

as in categorization, the trick for successfully performing this task is understanding something general about the situation instead of just something specific about the two objects in front of you. Understanding the overarching rules of a situation is something we are very good at. Indeed, eight-month-old infants are already capable of some forms of this ability to learn how to learn a task. We could call it "meta-learning."

Harlow's meta-learning experiment has now been done with many animals. Macaques do reasonably well, but squirrel monkeys do worse. They take about 1000 sets and only reach about 80% correct on the second try. The squirrel monkey is a lot better than the squirrel, though. Squirrels and other rodents such as rats and gerbils never get much above 60%, even after 1800 sets of trying.

Meta-learning and categorization are two types of behavior that rely on a distinct set of brain regions that non-primate mammals do not have, the temporal cortex and the prefrontal cortex. These regions are especially well developed in a specific group of primates to help them address specific foraging challenges. These are the simian primates.

———

The difficult life of a simian primate

Africa was traditionally thought to be the birthplace of simian primates, but that theory is disputed. What we do know is that the oldest primate fossils are from the northern continents, which now form North America and Eurasia. These fossils are from primates that appeared and rapidly diverged at the boundary of the Paleocene and Eocene geological epochs, about 56 million years ago. This time period was the warmest since the extinction event that killed off the dinosaurs and is termed the Paleocene-Eocene Thermal Maximum.

By the time of the PETM, primates had already split into two lineages that would become the ancestors of the two major groups of primates alive today. One lineage would become the ancestors of what today are mostly smaller primates character-ized by their wet nose—a bit like the nose of dogs and cats—long snouts, and comparatively small brains. This current group of primates are the strepsirrhines, consisting of the lorises of Asia, the bushbabies of Africa, and the lemurs of Madagascar. The present-day family members of the other group, the haplo-rhines, are generally bigger, dry-nosed, and have bigger brains. They consist of tarsiers and all primates we refer to as simian primates: the monkeys and apes. The strepsirrhine ring-tailed lemur and the haplorhine rhesus macaque whose brains we see in the figure at the beginning of this chapter look quite different. The lemur has a smaller body and wet-nosed snout that vaguely reminds of a cat, while the rhesus macaque with its bigger body, flatter face, and bigger brain feels much closer to ourselves. As the primate traditionally most often used in research, the ma-caque is often referred to as "the" monkey, the prototype of the diverse group of species.

At the time of the Paleocene-Eocene Thermal Maximum, the African and Arabian tectonic plates, together known as Afro-Arabia, were separated from Europe and Asia by the Tethys Seaway. As a result, few animals were able to travel to Africa from the northern continents, but some managed. In the early and middle Eocene there is evidence for strepsirrhine-like pri-mates in northern Africa, including some a bit larger than most, occupying a niche that later became more associated with sim-ian primates. The climate was probably warm, wet, and a bit seasonal. Plant life probably looked similar to that of modern-day tropical rainforests. Fruits, seeds, and leaves would be abundant. There is undisputed evidence for the presence of simian primates in Africa only in the late Eocene. Their origin is still debated, but one theory suggests that simian primates originated in Asia and then somehow found their way to northern Africa.

During the late Eocene and early part of the subsequent Oligocene, about 34 million years ago, the temperature on Earth fell drastically. The concentration of greenhouse gasses in the atmosphere plummeted, leading to a permanent glaciation of Antarctica. A possible contributing factor might have been that during that time Antarctica became isolated from South America and Australia, creating the Drake and Tasman passages. This enabled the formation of a circular water stream around the southern continent, now known as the Antarctic Circumpolar Current. This current keeps cold water around the pole at all times, preventing the mixing of polar cold water with warmer water from more northerly locations and lowering the surface temperature, creating a permanent fridge on our globe. The cold water from the Antarctic descends into the ocean and makes its way to the North Atlantic and North Pacific, affecting temperatures worldwide. A similar transport of cold water originates from the Arctic. Because of the resulting drop in temperatures, the Oligocene is considered a transitional period, when the earth moved from the tropical, warm climate of the Eocene to the colder, drier, more modern climate of the subsequent Miocene period. The Swiss paleontologist Hans Georg Stehlin called the transition to the Oligocene *la grande coupure*, the great break.

The northern continents saw a gradual reduction in the diversity of primates during this period. Northern parts of Afro-Arabia became drier, tropical forest made way for more open and savanna forest. However, tropical forests did persist in more southern locations. This gave the early strepsirrhines an escape. In contrast, the early simian primates in northern Afro-Asia adapted.

When the availability of food becomes less abundant, a primate can adapt by changing its body or its brain. This is beautifully shown by research by the primatologist Katharine Milton on New World monkeys. In a classic study, she compared spider and howler monkeys. Although these monkeys live in similar environments and descend from a common ancestor, their solutions to scarcity of preferred food are quite different. Both these simian primates are on a plant-based diet, eating fruits and leaves. Of these two sources of food, fruits are far more nutritious than leaves; however, fruit is not always easily available and the monkeys had to find ways to deal with this eventuality.

One way to adapt to such a situation is to adapt your body to be able to cope with a downturn in availability of your preferred food. You can grow a bigger body that requires less energy per unit of weight, meaning you can rely on foods of lesser quality than fruit. You can also adjust your digestive tract. This is what the howler monkeys did. Their colon is much longer than that of the spider monkey, allowing them to ferment plant fibers for a long time in their digestive system to extract more nutrition. The spider monkeys followed a different strategy. Instead of growing their digestive tract, they grew their brain. They became more clever in searching out fruit, even at times when it is only scarcely available. Their short digestive tract allows them to digest large amounts of fruit every day, even unripe, low-quality fruit if that is all that is available, to extract all the nutrition they need.

It cannot have been easy for a simian primate to forage in the deteriorating conditions of the Oligocene, after being used to the balmy, lush paradise of the preceding Eocene. If you rely on ripe fruit as your main source of food, your challenges are cut out for you. Fruit, of course, is seasonal, so not every type of fruit is available all the time. Fruit is distributed in patches—there are no natural orchards of monoculture fruits—meaning that the fruit that is in season might be located on trees far apart. In fact, they might be so far apart that you cannot see them and the distance you have to travel between patches might be so great that you cannot simply check them all out every single day. Your patches also diminish when you eat from them, so what was a good tree yesterday might be an empty tree today. The travel between patches can be not only long, but also dangerous by exposing you to predators. You can avoid predators by foraging only around dusk and dawn, but that makes your search for food even more time sensitive. And then there is competition to think about; competition from your conspecifics, but also from other animals interested in the same food.

The simians of the Oligocene probably used a combination of the strategies described by Katharine Milton. First, simian bodies became larger, evolving from the small 100–300 gram animals found in the middle Eocene to the 1–1.5 kilogram animals living in the Oligocene. They developed as daytime visual foragers, with comparatively smaller eyes that featured a sensitive hot spot with very high visual acuity, the fovea. Some developed an extra type of cone in their retina, allowing better color vision. With larger bodies also came proportionally larger brains. Some parts of their brain expanded more than would be expected based purely on the size of their body. These expansions did not happen all at once, but occurred step by step. Likely, the visual cortex expanded first. As we found in the previous chapter, the occipital cortex is the start of two parallel cortical pathways of association cortex processing visual information, namely the parietal cortex pathways processing vision for movement, and the temporal cortex pathways processing vision for perception. This latter pathway likely also expanded early in the simian adaptation.

Later, another part of the association cortex expanded: the frontal cortex. Before this happened, the simian primates had already diverged into two groups. One group stayed in the "old world" as the founders of the Old World monkeys and apes. Another group seems to have made the transition to South America, which started to separate from Africa about 140 million years ago. They founded the New World monkeys that now live in South and Central America. Independently of their Old World cousins, these New World simians also invested in an expanded brain. The frontal cortex expanded independently in both groups.

The newly expanded parts of their brain—the temporal and frontal cortexes—make modern simians masters at dealing with the kinds of challenges associated with a fruit-based diet. Observations of foraging simians in the wild show the wide range of information they can take into account. They can learn about specific trees, which ones produce high-quality fruit and how much. They can use a wide range of sensory cues to identify a source of good food, not only through high resolution vision due to their newly evolved ability to see three colors instead of two, but also through sounds made by their conspecifics and even by other animals interested in the same food. They can generalize across trees of the same species, meaning that if one tree provides good food at a particular time they are more likely to infer that others of the same type will as well. They can take the weather into account, inferring whether certain types of trees are worth visiting by taking into account how quickly fruit ripens in different temperatures.

But how exactly do these parts of the brain help to solve their foraging problems? This integration between neuroscience and primatology is largely the work of the neuroscientists Dick Passingham, Steve Wise, Betsy Murray, and their colleagues. They proposed a set of theories that explain how the simian expansions in association cortex evolved to help simians forage in a world where fruit became less abundant. Their theories explain how the simian macaque monkey's brain is much more complex than that of the strepsirrhine ring-tailed lemur.

———

Assessing a visual scene

How does our brain recognize a distant patch of promising fruit trees in a crowded rainforest full of visual stimulation and noise? Ideally, how does it recognize a nice patch that doesn't have too many jaguars who would like to forage on us? Come to think of it, how does our visual system recognize a piece of ripe fruit to begin with? This is the job of the temporal visual pathway, the second of the two visual pathways in the primate brain.

In the last chapter, we saw that visual information is not processed in our brain as a single whole. At least two visual pathways originate in the primary visual cortex at the back of our brain. We saw in patient V.K. that damage to the parietal pathway leads to difficulties in visually coordinated movements. In patient D.F., we saw that damage to the temporal pathway that runs behind the ear leads to a loss of conscious perception of objects, including ripe fruit. We concluded in the previous chapter that different types of visual information are processed in different parts of the brain, and this separation starts early in

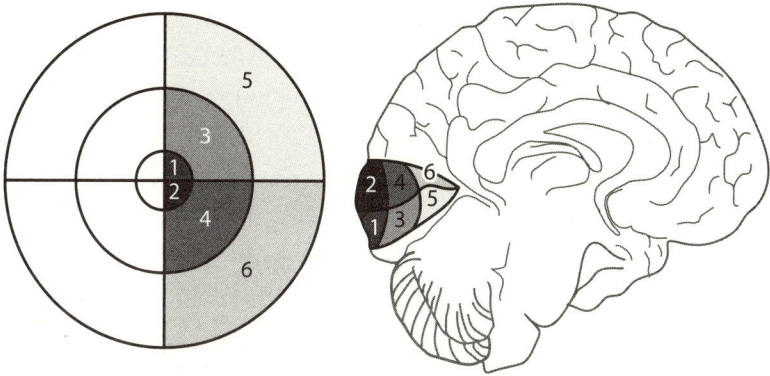

FIGURE 4.1. Retinotopy. The neurons of the early visual cortex are organized according to which part of the external world they represent following two overlapping principles: the polar angle (where along the 360 degree center of gaze) and the eccentricity (how far away from the center of gaze). Together, these principles mean that different parts of space are mapped onto different parts of the brain.

the brain's information processing pathways. Cells in the primary visual cortex do not respond to whole objects but to small features of a visual stimulus. They discriminate small differences in orientation of lines, the frequency of changes in the visual field, or changes in color. The primary visual cortex does not represent whole objects, but detects the small building blocks of visual stimuli. Another way of saying this is that the primary visual cortex detects visual features.

The feature detection cells in the visual cortex do not simply respond if a certain feature is present anywhere in the visual field. Instead, the visual cortex is organized as a map of your visual space. Different cells respond depending on which part of your retina is hit by the light from a stimulus. This map is organized along two dimensions that indicate the location of the stimulus. The first dimension captures the angle with respect to your center of gaze. In other words, if you imagine your visual field as a clock face, is the stimulus at two o'clock (top right) or three o'clock (center right)? In the first case, a specific group of visual cortex neurons will fire; in the second case, a group adjacent to it will fire.

The second dimension captures the distance from your center of gaze. In other words, are you looking directly at the stimulus or is it more in your periphery? Again, if we compare a stimulus close to our gaze with one slightly farther away, adjacent groups of cells will fire. Together, these two principles allow us to reconstruct where on the retina the light of a stimulus fell, based purely on which group of visual cortex cells fired. This principle is called retinotopy. Visual cortex contains feature detectors organized in a retinotopic fashion. From these elementary elements, the rest of the ventral visual pathway has to reconstruct greater wholes.

The visual cortex cells processing this early information send their information to cells along the temporal visual pathway. As we move more and more forward along the temporal visual pathway, the cells start to respond to more complex patterns. Rather than just responding to simple features such as lines and colors, they start responding to conjunctions of features such as shapes of a certain color. At the same time, their specificity to the location of visual information becomes less obvious. Thus, the cells start to represent more complex information, less dependent on the exact details of the visual stimulus. Ultimately, cells come to represent whole objects. By that time, the representation is quite distinct from the actual particular information from the retina. For instance, if we are looking at a banana, we will still be able to recognize it as a banana if it is oriented horizontally rather than vertically, or if it is partially hidden behind an apple. We will also recognize it as a banana whether it is peeled, unpeeled, on the green side, or beginning to look brown.

Some categories of stimuli tend to activate groups of cells close together. If we ask people to look at socially relevant stimuli like faces or bodies, we tend to see that distinct areas along the temporal pathway become active. Tools also tend to activate a distinct area in many people. These groups of cells responding to a particular group of stimuli have been termed "patches"—for instance, the human temporal cortex is thought to contain face patches, group of cells that fire when their owner sees a face.

Exactly how these different patches in the temporal cortex are organized and how they develop is still a matter of active debate among neuroscientists. Some argue for strict specialization of the different patches. Each patch is specialized for a particular type of stimulus and has a sharp border with a neighboring patch that is specialized for a different type of stimulus. Others argue that the patches are merely an epiphenomenon. Neurons responding to different stimuli are scattered around a part of the temporal cortex, and the patches are merely a reflection of slightly higher concentrations of neurons with different responses. The processing of information then results from the combined activation across a larger part of the temporal visual pathway.

One argument affecting these debates is how stable a particular patch's function is. Take a patch responding to faces, for instance. The suggestion by the original discoverers of these patches was that they are likely innate. As a social species, with babies fully dependent on their caregivers, it is a good evolutionary strategy to be able to recognize faces from the start. Hence, it was argued, face processing is probably innate. On the other hand, you could say that processing faces would likely be learned really quickly, even if it is not innate. Anyone who has a baby quickly realizes that most people—parents, siblings, grandparents, even total strangers in the street—can't help but hover over the baby's face and make funny noises at it. The baby quickly learns that recognizing that thing hovering above them and smiling at it leads to all kinds of desirable outcomes, including the provision of nutrition. Even if recognizing faces is not innate, it is likely one of the first things a baby learns.

We can leave the neuroscientists to debate the nature of the patches among themselves. For our purpose, the important point is that the temporal visual pathway allows simian primates to represent complex objects consisting of a combination of different features. It can do this independently of the size or orientation of the original stimulus, and it can even deal with partially obstructed stimuli or ambiguity in the visual information. The farther along the temporal visual pathway we go, the more different types of information—the more features—get integrated, and the more independent of precise information hitting the retina, the more abstract the information becomes. Information processing becomes less dependent on the precise details of the information processed in the primary visual cortex. Eventually, whether a neuron responds to a particular stimulus is no longer dependent on a particular feature or even a particular stimulus, but on distinct classes of stimuli. In other words, the responses of the neurons are dependent on the category a stimulus belongs to.

If we want to survive in a complex and ever-changing environment, it helps to be able to quickly evaluate what we see. Is this animal a dangerous one that I should avoid, or can I safely proceed? Is this fruit ripe and ready to be eaten, or will it give me a stomachache? Is this a type of tree that is likely to yield good fruit this time of year? Answering these questions means that we quickly have to *categorize* the visual stimuli we encounter.

This categorization often doesn't rely on any one particular feature of the stimulus, but on a combination of difficult to define aspects of the stimulus. For instance, there is no one thing that indicates whether fruit is ripe. Yes, color is one important thing, but there is no hard boundary between a ripe and an unripe fruit. The smoothness of the skin will give you another clue. The color of the leaves of the tree might provide a more contextual hint on whether its fruit is likely to be ripe. Even more contextually, the weather over the last couple of days might suggest that fruit is unlikely to be ripe at the moment, so that dark-colored fruit in front of us might be an anomaly. All this type of continuous information must be combined to give a best estimate of the ripeness of the fruit. There is no simple list of pluses and minuses; it is a complex integration of relations of different clues.

An experiment by the neurophysiologist David Freedman shows us how the temporal visual pathway can help achieve this. He gave macaque monkeys a task in which they had to identify stimuli as representing either a cat or a dog. The monkey was presented with one stimulus and then, after a delay while an empty screen was presented, a second one. If the two stimuli belonged to the same category, either cats or dogs, the monkey had to make a response. If they belonged to different categories, the monkey should not do anything and after a while a new stimulus was presented.

To solve the task successfully, the monkey had to recognize the category of each stimulus and be able to determine whether the category matched one that it had already seen. To make the task difficult, Freedman used a morphing software to create intermediate stimuli that were partway between a cat and a dog. So, he had some stimuli that were 100% cat, some that were 80% dog and 20% cat, some that were 60% dog, and so on. No single feature distinguished any stimulus as "cat" or "dog," but still if you see the stimuli some are clearly "cat-like" and others are clearly "dog-like."

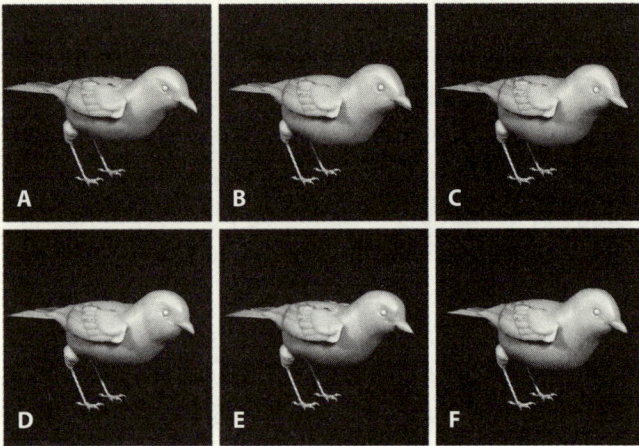

FIGURE 4.2. The type of stimuli used in the experiment where participants had to distinguish different categories of birds. Such experiments revealed how the temporal and frontal cortex work together to learn the categories and use them to guide behavior.

Neurons in the temporal visual pathway responded differently to the different stimuli. Particular neurons "preferred" a particular stimulus, exhibiting the strongest firing when it was presented. In most cases, they responded as well, but less strongly, to stimuli that were similar to it. In some of the neurons, the response profile was so clear that they effectively could be seen as distinguishing between the dog and cat categories. The area in the temporal pathway containing those neurons allowed the monkey to quickly analyze the various aspects of the visual shape of the stimuli.

When I was doing research for my PhD in the early 2000s, a colleague of mine, Marieke van der Linden, was doing similar experiments with humans. Interested in how we learn to distinguish different categories of stimuli, she used a software package to create images of six types of birds. The pictures were identical except that aspects of the birds' back, belly, tail, beak, head shape, cheeks, brow, and eye position could be manipulated. The differences were tiny; an untrained observer would be hard pressed to see the difference between the six birds. My colleague then used a morphing software to create hybrids between the different categories, just as Freedman had done with his dog/cat stimuli.

Human volunteers were trained for two hours per day on three consecutive days to distinguish the different bird categories. On the days before and after the three training days, the participants participated in an fMRI experiment in which they were presented with the different bird stimuli while their brain activation was recorded. Before training, brain activation to all bird stimuli was the same. After training, though, parts of the temporal visual pathway showed stronger activation to birds whose category the subjects had learned. This was not due to simple familiarity with the birds, because control birds the volunteers had seen equally often but had not learned to assign to a specific category did not activate the region.

Thus, the temporal visual pathway helps people categorize different bird types based on combinations of hard to describe and barely distinguishable visual features. The part of the temporal cortex that shows this effect is called the fusiform gyrus. It also contains a region that is important for faces—a face patch—which is therefore often called the "fusiform face area." In a way, the volunteers in my colleague's experiment now had a fusiform bird area. Presumably, the exposure to Formula 1 cars I mentioned at the beginning of the chapter means that over the years I might have acquired a fusiform Formula 1 car area.

The abstraction of the coded categories increases when we move more anterior in the temporal cortex. In humans, this part of the brain is expanded and is connected to many other parts of the brain. While information from the visual system and the auditory system stay relatively separate in the macaque monkey and the chimpanzee temporal lobe, in the human the connections bringing information from these systems overlap. It is thought that this is related to the emergence of a particular type of memory in the human brain, termed semantic memory. In contrast to episodic memory that contains information about your own life and experiences, semantic memory contains information about objective facts about the world.

For instance, my memory of going to kindergarten in the United States is an example of episodic memory, while my knowledge that kindergarten is a preparatory type of schooling before primary school is semantic memory. The contribution of the anterior temporal cortex to semantic memory places humans apart from other primates, but it is an elaboration on the neocortical system we see in all simian primates.

The importance of the temporal lobe in building categories is also seen dramatically in some neurological patients. Semantic dementia is a group of brain disorders often affecting the anterior part of the temporal cortex. Patients in whom this part of the brain is damaged often experience difficulty in describing boundaries between categories. They increasingly start to rely on superficial similarities to categorize stimuli, such as grouping all the yellow objects together, but not the banana and the apple or the sunflower and the rose. There are some suggestions that the kind of mistakes these patients make depends on the severity of their disease. Some tend to overgeneralize, calling everything a fruit, before they start to undergeneralize, failing to include fruits in their appropriate categories. Such distinctions, however, are hard to reliably observe due to the tremendous variation across individuals. But that the boundaries between categories become more blurred is clear.

At every step along the temporal visual pathway, visual information gets re-represented. Different features get combined and a more complete, more abstract picture gets built. Every re-representation allows the simian primates to better understand their environment. In foraging terminology, the temporal cortex evaluates the sensory context for foraging. The next step is for the animal to use this information to guide their foraging choices.

————

How the frontal cortex helps us forage

We all know the difficulty of not giving in to temptation. Sometimes we know something is bad for us—a bar of chocolate, another beer, a sugary cookie—but we want it anyway. Children in particular have difficulty delaying or denying themselves gratification. Ask a small child if she wants a cookie now or two cookies later and most of the time they will choose the cookie now. Or, as my children try to do, "negotiate" one cookie now and then two later. In the same vein, my three-year-old son likes to press any button he sees. Lights, oven, doesn't matter. Of course, he knows he is not supposed to press every button. Indeed, children can often repeat the rule they have been taught— "don't press the button"—at the same time they do it anyway. Clearly, their problem is not identifying the sensory context— they know perfectly well what is appropriate. Their problem is letting that information guide their behavior.

In the 1960s the Soviet neuropsychologist Luria described this phenomenon in patients with damage to parts of the neocortex. His patients were able to perfectly describe what they had to do in any given context. For instance, they were told a rule: "press the button that lights up, except when the background is blue, then don't press anything." Even though the patients were perfectly able to describe the rule, they still pressed the button that lit up, even when the context—the blue background—told them not to. They understood the context, but they could not translate that to their behavior. It was as if their conceptual system and their motor system were disconnected.

For a foraging simian primate, it is important to know the sensory context, but it is also important to know what to do in that sensory context. And that might not always be the same. Something that looks like ripe fruit might not be good to eat in certain contexts, like when it is hanging from a tree whose fruit previously made us sick, or simply when we are not hungry. The context might also change very quickly; what was a good thing to do yesterday might not be a good idea today. The foraging brain thus needs a mechanism to transform the context into suitable behavior.

Transforming contexts into suitable behaviors is precisely what Luria's patients could not do. The temporal cortex cannot do this for us—it has no direct connections to any of the brain areas influencing the outside world. As we saw in the previous chapter, the "motor cortex" is located in front of the big, central groove in our neocortex. The motor cortex does not receive any direct information from the temporal cortex. For information from the temporal cortex to reach the motor cortex of the brain, the information has to go via the part of the brain in front of the motor areas. These areas are located at the most anterior end of our brain, hence they are called the prefrontal cortex. It was the prefrontal cortex that was damaged in Luria's patients.

To compare the contributions of the temporal and the frontal cortex, let's look at the monkey experiment of David Freedman again. When he recorded from the temporal cortex, he found neurons that signaled different dog-like or cat-like stimuli. Subsequently, he recorded from the frontal cortex while monkeys responded to the same categories. To make the task work, the monkey had to not just look at the stimuli but also remember them over a short interval. They then saw another stimulus, and when that stimulus was of the same category as the first ("a match"), the monkey had to make a response. If the categories were different ("a non-match"), the monkey had to do nothing. Thus, the monkey had to remember the category it had seen during the interval and compare that with the category of the next stimulus and decide whether to make a response or not.

This is quite a complicated task. Although the temporal cortex is helpful in identifying the stimulus categories, this is not enough to solve the task. The recordings from neurons in the frontal cortex showed a very different pattern from those we saw earlier in the temporal cortex. From the beginning, the neurons in the frontal cortex showed a firing pattern more clearly related to what was important to solve the task: the stimulus categories. While you could reconstruct which individual dog or cat the monkey had been looking at from neurons in the temporal cortex, the frontal cortex neurons really just coded the category of a stimulus.

Importantly, the frontal neurons kept firing during the time the category had to be remembered, even though temporal cortex neurons were silent. Then, when the next stimulus was presented, some neurons in the prefrontal cortex still kept responding to the previous category, precisely the information the monkey needed at that time to make its match or non-match decision. It seemed that, while the temporal cortex was able to provide individual pieces of evidence needed to solve the task, the frontal cortex was the place where the different pieces were combined into a coherent whole.

Simian primates have a series of regions within their prefrontal cortex that other animals, including strepsirrhine primates such as the lemur, do not have. These regions have widespread connections with the rest of the brain. Together, they can combine information from different reference frames, different types of sensory modalities, even different time intervals. It is often said this part of the frontal cortex, termed the "prefrontal cortex" because its regions are located toward the front of the brain, forms the top of the "hierarchy" of the brain. This is dangerous terminology, because it makes it sound like the prefrontal cortex is an all-important controller in the brain. But of course, there is no single controller. Who would be controlling the controller?

The prefrontal cortex receives information from many other parts of the association cortex, providing it with a wide range of information. It receives information about the sensory context from the temporal cortex, but also information about space and the body from the parietal visual pathway, about the internal state of the organism, about the body's state, about past memories, and much more. Together, the different parts of the prefrontal cortex are placed best to integrate all the information available and decide how best to organize behavior. In an analogy with our story of feature combinations in the temporal cortex, we can say that the prefrontal cortex can combine different information representations from various parts of the brain. Prefrontal areas literally take the representation of information to a new level. In foraging terminology, the prefrontal cortex has all the necessary information to be able to establish goals.

There is another important feature of this prefrontal role in establishing goals. In the case of David Freedman's categorization experiment, the prefrontal cortex uses information from the temporal lobe for its task. But it is also capable of helping the temporal lobe. If we change the categories in the experiment, the monkey can learn the new categories. It can even do that if it has no frontal lobe. As we said, the temporal cortex can learn the categories, but it only does so really slowly. It will take a monkey without a frontal cortex a very long time to learn. A monkey with a frontal cortex, however, can learn the task again quite quickly.

Having a prefrontal cortex helps speed up the learning process by the temporal cortex. This is what makes the frontal cortex such a powerful tool for simian primates. In their uncertain world, they need to combine a lot of information from different sources. The prefrontal cortex can help with that, but it also helps them to learn fast. In a fast-changing world, a foraging animal cannot perform hundreds of trials to learn something the way it can in a laboratory. It needs a good strategy quickly. The prefrontal cortex, with its goal representations based on integration of all sorts of different information, helps with that. This is what we see in the experiment by Harlow at the beginning of the chapter. Remember that his task was given to many different animals, and their ability to perform the task differed among them. The animals that learned the task more quickly shared one thing: a bigger prefrontal cortex.

Still, it cannot be emphasized enough that the prefrontal cortex is not an all-mighty controller. There is no one place in the prefrontal cortex where all the information comes together. Rather, different parts of the prefrontal cortex combine different sorts of information. Different parts of the prefrontal cortex interact with one another, and eventually a consensus emerges. Many neuroscientists have likened this aspect of brain function to an orchestra without a conductor. All the instruments play and a symphony emerges, but there is no one person in charge.

Recap: The lemur and the macaque

Areas such as the temporal cortex and prefrontal cortex expanded significantly in simian primates and have kept expanding in some groups. They are expensive to maintain, so they only expand when they give the owner an advantage in foraging for their favorite food—the not always available, patchily distributed ripe fruit. In some groups these areas even shrank again, presumably when their cost outweighed their advantage.

Because of the fascinating behaviors seen in some patients with prefrontal lesions and its location all the way at the front of our brain, the prefrontal cortex in particular has always fascinated researchers. Debates on whether the prefrontal cortex is particularly big in humans—potentially accounting for what people see as our more sophisticated behavior—rage in the scientific literature until this day. But the prefrontal cortex is nothing without the other parts of the brain, some of which also dramatically expanded in some primates.

What matters is that these regions appeared in a certain group of animals at a certain point in their evolution when the animals needed to solve specific foraging challenges. The abilities afforded by these brain regions helped them do that by being able to integrate a lot of information, and to do so quickly. It is what allows us to quickly learn and adapt to new circumstances, as in the learning to learn task we described at the beginning of this chapter.

The simian primates did one more thing to help their foraging. They became more social. Many of them live in large groups. This has often been taken as evidence against the idea that their expanded brain areas contribute to foraging behavior. But that is a false dichotomy. Indeed, in her research about how New World monkeys deal with food scarcity, Katharine Milton pointed out that the cleverer animals also use a social strategy by teaming up with others. It turns out, many of the areas that provide benefits in foraging also help to make you socially smart. That is the story of the next chapter.

5

The fox and the dog

THE DOMESTICATED, SOCIAL BRAIN

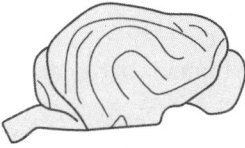

Red fox

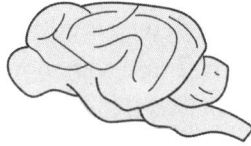

Domestic dog

There is an experiment I like to do with my cat. He has been with us for more than a decade, but because we got him from the shelter where lived all alone until he was four months old—nobody wanted a black kitten, superstitions die hard—he has never lost his fear of humans. If I move my head toward him too quickly, he will get scared and run away. But I noticed that if I move my head toward him while looking in another direction, he tolerates it. Apparently, he can distinguish between situations where my eyes are looking at him and when they are not. Not bad for a cat that can't figure out a route indoors when the open French doors block his cat flap.

It turns out I was not the first to come up with this experiment. In the early 1990s, a series of similar studies was done using black iguanas at a field station in Costa Rica. Humans sometimes eat iguanas in Central America, so they can be considered one of its predators. In one experiment, an experimenter either walked toward an iguana while looking at it, walked toward it while not looking at it, walked past it while looking at it, or walked past it while not looking at it. The iguanas fled most quickly from the approach with direct gaze, followed by direct approach with averted gaze and tangential approach with direct gaze, and least from the tangential approach with averted gaze. Importantly, the behavior of the iguanas was determined in part by their previous exposure to humans. Those with more exposure to humans better discriminated between the direct and tangential approaches.

In other experiments, it was tested what aspects of the eyes the iguana noticed. They responded faster to large compared with small eyes drawn on masks worn by the experimenter, indicating they pay attention to the eyes. If the response to a human walking toward the iguana was compared to the response to a human whose face was covered with hair, which was done to create the illusion of a person retreating, again the iguanas responded less than to the normal, approaching face. Similar sensitivity to being looked at has been shown in plovers, chickens, sparrows, green anole lizards, and, if we include the informal setting of my living room, cats. Whether one is being looked at is thus something that can be detected by a wide variety of animal species. So much so that some butterflies have giant eye-like motifs on their wings to fend off predators. It's a good evolutionary strategy to be able to detect whether a pair of eyes is looking at you, because if it is, it usually means you're a candidate for dinner.

We can get more information from eyes than just whether they are staring at us. Eyes help us recognize individuals and are crucial for recognizing facial expressions. Following another person's eye gaze is a useful skill. It might tell you a lot about their view of the world. If somebody looks at food you're interested in, it likely means you're in for a contest. So eyes can be very useful in social situations.

Sure enough, if we track the eye movements of macaque monkeys while they view pictures of other monkeys, they tend to focus mostly on the eyes. The same happens when they view faces of apes, humans, or even schematic faces drawings. And that is not all. For humans, eyes are the "windows to the soul," which is an unscientific way of saying that we use the information we get from people's eyes—whether it's identity, emotion, or direction of gaze—to construct complicated ideas about other people's thoughts, beliefs, and desires. This has been termed "mentalizing," the ability to imagine what's in the mind of another and to use that information in social encounters. This is something so important for human society that an inability to mentalize, as seen in some developmental conditions such as autism, can severely impact a person's life.

Although paying attention to eye movement to check if you're a candidate for dinner is a widespread ability in the animal kingdom, the more complex abilities do not come easily to all animals. For example, it remains unclear how much information about another's intentions nonhuman primates can infer from following eye gaze. Initially, it was thought that chimpanzees could not make inferences about the mental state of another being—they were thought not to be able to mentalize. In one early experiment, chimpanzees were presented with two food containers, only one of which actually contained food, and the experimenters wanted to see whether chimpanzees could use social cues to find out which was which. The experimenter looked at the container with food or even pointed at it, to indicate to the chimpanzee which one to choose. The chimpanzees, however, performed at chance level, suggesting they did not use any social information to guide their choice.

But then the anthropologist Brian Hare devised an experiment that showed that chimpanzees actually do infer a lot from eye gaze. They placed a chimpanzee opposite a dominant conspecific. In between the two animals there were food rewards, some of which the dominant could see, some of which it could not. As any chimpanzee knows, stealing food that a dominant has his eye on is not a recommended strategy. In Hare's experiment, the subordinate chimpanzee subject used the information about which food the dominant animal could and could not see to plan their own move. They went only for the food the dominant could not see. In other words, they were able to infer what the dominant chimpanzee knew based on his gaze. Comparing this experiment with the situation where the chimpanzee ignored the helpful gazes of its caretaker, it turns out chimpanzees make social inferences in specific circumstances: they understand competition with a dominant chimpanzee, but not the cooperation of a helping hand. This fits with chimpanzee society, one of competition rather than cooperation.

Living in groups can have great advantages for foraging animals. There is safety in numbers when you forage in open environments where you are likely to be spotted by predators. There is also more knowledge in groups, as evident by the fact that foraging time decreases in larger groups, where individuals can exchange more information about where scarce resources are located. Teaming up with others can thus be a good foraging strategy. But living in groups is complicated. If you live with more conspecifics you are surrounded by individuals that want the same food as you or, worse, the same sexual partners as you. If you want to enjoy the benefits of social life, but not lose out against others like you, you need to have specific skills for navigating the social world.

As the chimpanzee example demonstrates, such skills are not a given. Could it be, then, that social information processing is just another possible adaptation to help solve the problem of foraging? Do the brains of some animals contain adaptations for processing social information, just as they have adaptations for other foraging skills, such as navigational memory or complex visuomotor skills? Are all social animals the same, or do their social abilities depend on their particular needs? To study these questions, we can again compare the brains and behaviors of related animals that evolved in slightly different niches.

There was another animal that Hare and his colleagues tested to see how it fares in the cooperative version of the gaze-following experiment. While the chimpanzee failed to take into account the helpful information provided by the direction of gaze of the caretaker, this one animal was able to use this information to its advantage. This animal was not a close primate relative of the human, but rather its long-term companion: the dog.

Dogs are domesticated wolves. Where wolves tend to shy away from humans and not interact with them, dogs love being with humans and can interact with us in ways no other animal can. Through the process of domestication, they acquired social skills that the chimpanzees never did. Can domestication provide us a first case to understand how a brain becomes social?

––––––

How to domesticate a brain

Dmitri Belyaev had a theory. As a lead scientist of the Department of Fur Animal Breeding at the Central Research Laboratory in Moscow in the late 1940s, he was tasked with improving the quality of fur bred from foxes and minks in the Stalinist

Soviet Union. The export of high-quality fur was one of the few ways available to the Soviet government to bring foreign currency into the country. The laboratory understood the techniques of selective breeding quite well, based on the knowledge of breeding domesticated animals that humans have been acquiring since the agricultural revolution. But how the early process of domestication of animals itself got started remained a mystery lost in the mist of time.

One thing that intrigued Belyaev was how similar types of changes occur in many different domesticated animals, even if it is unlikely they are selected for. For instance, most domesticated animals develop patching of different colorings on their fur and hides, such as the black and white spots of cows. A youthful appearance that stays in adulthood, including floppy ears, babyish faces, and curly tails are other traits seen in many domesticates. More important for the breeding industry, domesticated animals breed year-round, whereas their wild counterparts often adapt their timing of birth to the time of year when food is most abundant.

Belyaev also noticed that some of the characteristics in domesticated animals, such as the spots on the fur, appeared in some generations, only to disappear in the next generation and then reappear in the grandchildren. It was as if the animals possessed genes for these characteristics that were switched on in some generations. Belyaev thought these characteristics were not independently selected for by their breeders. Why, after all, would a pig farmer be interested in a curly tail? Belyaev reasoned that the unselected domesticated characteristics must be a consequence of active selection for a particular characteristic that the breeder does find particularly desirable. He decided to test his idea using the foxes at his fur breeding farms.

Domestication is a useful way to study evolution. It creates a clear ecological context to which the animal must adapt, namely a niche of dependence on and usefulness to humans who control the animal's reproduction and food supply. Moreover, the process of selecting for certain characteristics by selective breeding is one that has been well understood for a long time. Darwin himself thought selective breeding was familiar enough to use it as a first example when outlining his theory of natural selection in his 1859 book *The Origin of Species*. In fact, *The Origin* was only meant as a summary of his theory, and he later devoted a two-volume work to animals and plants under domestication. He used it to demonstrate the idea that there is natural variation of a characteristic in a population and that if selection—in this case human preference—means that individuals with a certain characteristic have more offspring, the characteristic becomes dominant. Darwin had already noticed that domesticated animals shared many phenotypic changes. Darwin wrote before the mechanisms of inheritance were known, and the volumes on domestication present his ideas on what this mechanism might be.

Domestication of animals—and plants, but those are not our immediate concern here—has been tremendously useful to humans. Dogs provided protection and helped when hunting. Goats, cattle, and pigs have been kept for food and later for dairy. Horses were useful for their muscles as riding animals and, along with cattle, to pull the plough. In fact, it seems surprising that so few animals have been domesticated. According to the biogeographer Jared Diamond, this is because domestication of animals relies on a series of characteristics that most animals don't possess. For domestication to be viable, we humans have to be able to provide the animals with their food (which means we have no domesticated anteaters), the birth rate has to be acceptable to humans (no domesticated elephants), and the animals have to want to breed in captivity (no domesticated pandas, as many zoos will testify).

Some other important characteristics are social. It's helpful if the to-be-domesticated animal is already a social animal, ideally with a strong group hierarchy. This is evident in cows, where the farmer only needs to lead the dominant animal to the barn in the evening and the rest will follow. Even more important is the animal's tolerance of humans. Zebras, apparently, are vicious animals that don't tolerate humans at all. According to Diamond, zebras injure more zookeepers than tigers do. This was bad news for sub-Saharan Africa, which could not exploit the horse-like power potentially provided by zebras.

Gazelle also don't domesticate well, because they are extremely panicky around humans. Tameness, the willingness to let humans approach without attacking them or fleeing does seem to be a crucial characteristic to select for if you want to domesticate an animal. This is precisely what Belyaev planned to do with his foxes. Tameness is a characteristic that most breeders like to select in their animals. After all, for most animals domesticated by humans, whether cows, 1000 pound pigs, or dogs that could easily bite our hand off, tame specimens would be quite preferable to work with.

Belyaev decided to study how tameness and other characteristics of domestication were related by selectively breeding the foxes most tolerant to humans. He had to start his experiment with the help of breeders at farms as far away as possible from Moscow, as genetic research was regarded as highly suspicious by the Stalinist regime. He convinced a trusted breeder in the borough of Kohila in Estonia to devise a simple test to assess the tameness of each fox. They slowly approached the cages, opened the cage slowly, and reached in with one (gloved) hand containing some food. Most foxes would back away and react aggressively, but a few were a little less distressed. The tamest foxes were then selected and allowed to breed. From this generation, again the tamest foxes were allowed to breed. And so on. After as little as three breeding seasons, some of the fox pups would already react with less aggression than their parents and grandparents, an extremely fast pace, evolutionarily speaking. The pilot project seemed successful.

By the late 1950s political winds in the Soviet Union had begun to shift, and genetics research was no longer the pariah it once was. Belyaev became head of the new Institute of Cytology and Genetics near Novosibirsk. Finally, he and his collaborator Lyudmila Trut were able to expand the pilot study into the experiment he had always dreamed about. The results were nothing less than astonishing. After a few generations, Trut and Belyaev noticed that some of the tamer female foxes went into estrous a few days earlier than normal for silver foxes and produced on average slightly larger litters. A key domestic trait was starting to emerge within the foxes. With generations continuing, more and more traits started to emerge.

Usually this happened to very few animals at first, but eventually the traits would become more prevalent in the tame population. The fox pups started to actively seek out human attention, completely opposite from the normal fox behavior of shying away from humans. They started to whine when the humans ignored them, lick their hands when fed, and—starting with a little pup called Ember—wag their tail when Trut approached. In short, they started to behave more like dogs than foxes. Their appearance changed too. Their fur became patterned, the faces of the adults retained the juvenile shorter snouts, and their ears became floppy. In another crucial development, the foxes started to allow a person to stare at them, clearly seeing the human as a companion rather than a threat. All of this happened just following selection for tameness, as a package deal. Nowadays, this phenomenon is called "domestication syndrome."

The domesticated foxes were clearly more comfortable with humans. Indeed, they craved human companionship as much as a dog. But were they also better at reading humans than wild foxes? Brian Hare, whom we met earlier testing dogs and chimpanzees, worked together with Lyudmila Trut to test whether the domesticated foxes could solve the pointing and gaze-following task devised for the dogs. The tame foxes passed the tests with flying colors, touching objects the human experimenter gazed at or touched more often than the untamed foxes. The fox pups weren't different from dog pups in their ability to perform the task. Clearly, even though they were only selected for tameness, their ability and willingness to follow cues provided by humans was greatly enhanced. For dogs, you could argue that their tameness is the result of living their whole life amongst humans, but this was not the case for the foxes who were raised without much contact with their human caretakers.

How can selection for a single trait cause so many changes? Come to think of it, how can the same set of changes occur in such different domesticates, even across animals from very different branches of the mammalian evolutionary tree? Belyaev observed that the changes occurred too fast to be due purely to genetic mutations. He proposed the idea of "destabilizing selection." The variations associated with domestication might be present for a large part in the wild animals, but are not activated. When the selection pressure changes due to the influence of humans, these latent variations come to the surface. Not all genes are turned on all the time. Nowadays, the idea that genes do not produce the proteins that influence processes in the body all the time is well established. This partly explains why not all animals can be domesticated, since the wild animal already has to have the appropriate latent characteristics. It also explains the speed of the changes and the fact that certain characteristics can appear, disappear, and reappear across generations.

But what do these genes code for? One recent hypothesis that tries to account for the varied changes in domesticated animals is the "neural crest hypothesis," which attributes an important role to stem cells. Stem cells are cells that have not yet specialized into cells with a specific function, such as liver cells, skin cells, or neurons. Neural crest stem cells are specific to vertebrates as they form along the neural tube that later during development turns into the brain and spinal cord. During development, neural crest cells go on to specialize and influence tissue in the skull, the cartilage of the tail, and pigmentation.

As we have seen above, these are all parts of the body that are influenced in domestication, including a more juvenile skull, a shorter and curly tail, and more patchy pigmentation. Although the neural crest cells don't directly influence the brain, they can influence the gland producing adrenaline, the hormone responsible for the fight-or-flight response. Adrenaline glands are reduced in tame animals, leading to less jumpy behavior and, as a consequence, a greater chance to get used to interactions with humans. The neural crest hypothesis suggests that the selection for tameness results in a reduced number of neural crest cells at the sites of the body where they specialize. This could happen through different genetic mechanisms in different species but will have similar results—similar phenotypes in genetics-speak.

But what is different about the brains of domesticated animals? The neuroscientist Dieter Kruska spent a lot of his career comparing the brains of wild and domesticated mammals. He found that the brains of domesticated animals are consistently smaller than those of their wild relatives, even when taking into account the reduced body size of the domesticates. But the story is more complicated than a simple overall brain decrease. Just as increases in brain size are usually not uniform throughout the brain, the reduction in brain size also targeted specific parts of the brains.

The precise details differ across species, but it is the telencephalon that generally shows the greatest size decrease. Recall from chapter 1 that the telencephalon is the front part of the vertebrate brain, which often changes a lot in response to new foraging challenges. Within the telencephalon, it is particularly the neocortex that is most reduced in size. One hypothesis is that this reduction in the neocortex in particular is due to the reduced need for domesticated animals to devote much brain power to complex cognitive processes associated with foraging, since all their needs are seen to by their human caregivers. Consistent with this idea, some experiments have been conducted showing that domestic dogs might not be as good at problem solving as wolves, although superior learning and memory capabilities have also been found in domesticates. In the case of dogs, some of their worse performances on problem solving tasks might be due to the dogs relying more on human encouragement in these situations and giving up more easily when on their own, suggesting that their poor performance is not due to a lack of intelligence per se. Complicating the situation, brain size reduction has been found in most domesticated animals but not in the Russian foxes.

If the neural crest hypothesis is true, it should be able to account for these changes. Neural crest cells do not turn into brain cells, but they do seem to be involved in the chain of events that lead to the development of the telencephalon. Animals in which the neural crest cells are experimentally removed fail to develop major parts of the brain. As such, the hypothesis can go some way in explaining brain differences between wild and domesticated animals, especially the early stages of domestication, when a reduced stress reaction to humans is the necessary change that enables interaction between species in the first place.

The question, though, is whether just this general reduction of brain size can explain all the behaviors of domesticates—particularly the behaviors in which they excel, including their social skills. One idea consistent with the neural crest hypothesis is that domesticated animals in a sense stay more "youthful" during their adult life. Their appearance, with their short snouts, indeed looks more youthful. It could potentially account for increased social abilities. Many animals have a socialization window, a time during their development when they are more tolerant of others and more able to learn from them. Could it be that the increased social abilities are not so much acquired in domesticated animals as that they are just not lost when they become adults? That, together with a recognition that the human is a close companion, would be able to account for some of their social abilities.

But obviously not all domesticated animals excel at the same things. We domesticated the wolf to be our companion, but the same cannot be said for the cow, sheep, or even the horse. Even different dog breeds differ in their useful abilities. The Harvard neuroscientist Erin Hecht tested whether these different abilities were reflected in the brains of the different breeds. Using MRI, she scanned the brains of dogs of different breeds and investigated the relative size of different networks of brain regions. She found that breeds selected for scent hunting showed an increase in the volume of areas processing smell, whereas sight hunters showed an increase in areas controlling eye movement and mediating spatial navigation. In contrast, a network of regions that Hecht and colleagues identified as likely involved in social interaction seems particularly prominent in, among others, the Yorkshire terrier lap dog.

This work suggests that becoming a social species is not a one-step process. Selection for tameness is a crucial first step, and this will lead to changes in the domesticate's brain and behavior, including perhaps an increased sensitivity to processing social information as a carryover from youth. After this first step, however, increased selection will lead to further changes in the brain, favoring the brain regions important for the behaviors of interest. Understanding this second step is crucial if we want to understand the advanced social skills of primates, including humans.

————

The complexity of primate social life

Primates as a whole are social. One hypothesis explaining at least some of this sociality is that primates are "self-domesticated," meaning that they evolved to exist more in harmony with other members of their species. The idea is that learning to be close to conspecifics set in motion the same types of changes that occurred in dogs and in Belyaev's foxes. A particularly strong case has been made for our closest relatives. The chimpanzee genus *Pan* consists of chimpanzees and a sister species, the bonobo. Chimpanzees and bonobos diverged some two million years ago, long after the split between the *Pan* lineage and the one leading to humans.

As bonobos live south and chimpanzees north of the Congo river, it is thought that the formation of this large waterway isolated two groups of chimpanzee-bonobo ancestors that subsequently diverged. The territory south of the river affords large patch sizes and high-quality food. In contrast, in the north, food availability is reduced and there is also competition for feeding resources from gorillas.

Chimpanzee males use aggression against members of their own group when competing for dominance and in sexually coercing females, and against neighboring groups when defending their territory. Bonobos, in contrast, are known as the "hippy apes," where females tend to have food priority. They rarely use physical aggression and seem to use sex in every possible partner combination as a form of social grooming and tension reduction in the group. Incidentally, the frequent sexual contact of bonobos is one reason we know less about them than about chimpanzees, as it was long considered inappropriate to have animals behave in such a way in reputable zoos.

The reduced aggression in bonobos has led Hare and colleagues to suggest that this species underwent self-domestication. One theory is that the reduced food competition south of the Congo River enabled females to form coalitions against the aggressive males, which in turn led to less aggressive males having more offspring and setting the selection for tameness in motion. Just as with domestication in dogs and foxes, reduced aggression led to a host of changes in the animal. Bonobos' physical appearance is indeed more juvenile than that of chimpanzees, with reduced facial projections, smaller canine teeth, and a reduced cranium size. The skull is particularly interesting, as its shape is more similar to that of a juvenile chimpanzee, suggesting that it is the bonobo that changed to becoming tamer with respect to the chimpanzee-bonobo common ancestor, rather than the chimpanzee becoming more aggressive.

Some differences between the brains of chimpanzees and bonobos have been found, with bonobos having proportionally smaller areas in parts of the prefrontal cortex and a smaller hippocampus. James Rilling of Emory University investigated the connections of chimpanzee and bonobo brains, reporting an increase in the bonobo's connections between prefrontal cortex and the amygdala, a region in the brain mediating, among others, fear responses. He interpreted this as consistent with the idea that bonobos can better mediate aggressive responses and have more empathy in their interactions with others.

But primate social life is not just about domestication. What makes primates social is not simply that they live in large groups where they need to tolerate one another—primate social life is also highly complex, and primate brains have evolved to create and navigate that complexity.

Just how complex primate life is was demonstrated in the 1980s by the primatologist Frans de Waal. He observed a colony of chimpanzees in Burgers Zoo in the Netherlands, detailing how different individuals play different roles in the group. Chimpanzees, just like many other primates, form dominant hierarchies where one animal has preferential access to food. The common assumption was that this was survival of the fittest, the strongest animal occupied the top spot in the hierarchy. De Waal saw something different. He observed how two subordinate animals teamed up, forming a coalition to dispose of the dominant.

Other animals played particular social roles in the group. An older female called Mama acted as the matriarch of the group. She was the arbiter in conflicts between the males and made sure animals who got in a fight eventually made up. One particularly aggressive male had to be removed from the group, as his presence became a disturbance. When De Waal showed the other animals a video of this male a year later, they still panicked. Clearly, the individual and his actions had not been forgotten. The complex interactions involving coalitions, making up after a fight, and long memories of the actions of individuals that De Waal showed in his observations gave us a new appreciation of the complexity of the social lives of our closest relatives.

The complexity of primate social relationships was recently demonstrated in dramatic fashion in a natural experiment caused by an environmental disaster. On September 20th, 2017, the category 4 hurricane Maria reached the little island of Cayo Santiago just off the coast of Puerto Rico. Just like the main island of Puerto Rico, Cayo Santiago suffered widespread devastation among its inhabitants. The inhabitants of Cayo, though, are not humans but macaque monkeys. The island is a research center of the American National Institutes of Health and the University of Puerto Rico. It houses a free-ranging colony of macaque monkeys whose behavior and cognition are studied. The researchers have established detailed records on the social lives of all the animals.

As the monkeys are provisioned with food and water, the death toll in the aftermath of the hurricane was relatively minor. This allowed the researchers to see how the social fabric of the community changed following the disaster. Animals tended to spend more time in close proximity to others and spend more time grooming others, especially if they had been less social before the disaster. Interestingly, rather than focusing more intensely on existing relationships, the monkeys tended to increase the number of individuals they interacted with. They especially reciprocated grooming from others and closed relationship triads, becoming friends with their friends' friends.

In other words, the monkeys significantly increased their social networks in the aftermath of a natural disaster and did so in a way that was predictable, making friends with those already within close social reach. Forming many relationships proved useful for the animals, as the massive defoliation of the island due to the hurricane meant there were few places to hide from the heat, which made it necessary for the monkeys to spend more time in close proximity to others. Strong, intense relationships are not useful when searching for precious space, but many weak relationships are. The monkeys thus restructured their social network to help them cope with the effects of the natural disaster.

These complex social behaviors in primates illustrate that not all group living is the same. A herd of wildebeest, although consisting of many individuals, does not have the complex social system of primates. Emphasizing the complexity of primate societies, the evolutionary anthropologist Robin Dunbar has shown how the social systems of primates differ from those in other mammalian species and in birds. In primate societies, individuals have specific, one-to-one relationships with other individuals. Maintaining such dyadic relationships requires recognizing each other, remembering the past history of interactions, having an emotional bond with the other, and maintaining the relationship through time-consuming social grooming. These dyadic relationships are embedded within complex networks of relationships, in which you not only have to be aware of your relationships with others, but also the relationships the others have among themselves.

The complexity of social life also presents a trap, since coherence of the group and reliance on others means that an animal cannot simply leave or join any group to suit their current fancy, as wildebeests can easily do. Acknowledging the complexity of primate societies, Dunbar proposed the "social brain hypothesis," which states that social life is so complex that the size of the brain of a primate presents a "ceiling" to the complexity of the social system of its owner. Dunbar and his colleagues have shown that neocortex size across primates varies with many variables describing the complexity of their social lives, including the size of the social group, the size of the clique of animals grooming one another, and occurrence of coalition forming. Outside of primates, such a clear relationship between brain size and a social structure is generally seen only in animals that form monogamous pair-bonds. Apparently, married life is complicated.

Social brain areas

The relationship between brain size and social complexity is interesting, but it does not tell us which specific aspects of neocortex organization accompanied the evolution of complex social abilities. In order to understand how specific parts of the brain evolved to support sociality, we need to explore what types of social information they process.

One way to investigate which parts of the brain are important for a given type of behavior is to look at individual variation in that behavior across individuals. It has been known for a long time that training in a certain task will lead to changes in the brain areas important for that behavior. One of the most famous examples of this type of study is the London taxi driver study. London taxi drivers are trained for up to two years in navigating London's many twisty-turny streets before they are allowed to drive passengers around—they are turned into expert navigators.

In 2000, Eleanor Maguire and her colleagues scanned the brains of taxi drivers and found that the hippocampus, a part of the telencephalon important in memory, was bigger in taxi drivers compared with nondrivers and that this increase was predicted by how long they had been in the job. This result pleasingly dovetailed with observations of differences in hippocampal size between animals that engage in behaviors that require a good spatial memory such as food storing and those who don't, and with seasonal differences in hippocampus size between the food-storing season and other times of the year in chickadees. Similar experiments have now been done for many behavioral domains. I participated in a study where I had to learn to juggle for six weeks, which led to a detectable difference in my posterior parietal cortex, consistent with the visual-motor role discussed in chapter 3. Fortunately or unfortunately, I then was not allowed to juggle for four weeks, after which the changes in parietal cortex were gone again.

This type of manipulation could in principle be applied to study brain areas important for social skills. It is of course more difficult to manipulate people's social life, but by chance such an opportunity did present itself in macaques. At some point, almost all macaque monkeys in a research center in Oxford underwent brain scans in the context of the various experiments they were involved in. Macaques are housed socially, in groups of varying sizes. This is best for these animals, who as primates are very social by nature.

Just as they would in the wild, the groups immediately form social dominance hierarchies. The group sizes and membership themselves are determined by the monkeys' caretakers. Thus, unintentionally, this led to a manipulation of the social complexity—as indexed by the number of group members—of the animals' lives. Using the same technique as used by Eleanor Maguire and her colleagues for the London taxi drivers, Jerome Sallet and his colleagues investigated whether there were areas of the macaque brain showing increases in size as a function of the group size of the animals. This analysis showed a part of the cingulate cortex and another area in the prefrontal cortex, but mostly a series of regions along the temporal cortex. Most prominent was a region in the middle part of the temporal cortex.

As we established in the previous chapter, the temporal cortex contains areas that process the two modalities that best help one get information from a distance: vision, through the temporal visual pathway, and hearing. Turns out, these are also the modalities that primates use for acquiring a lot of their social information. Yes, smell and touch are important in our social life, but the primate specialties in processing social information rely on sound and above all vision. The temporal visual pathway contains a number of areas specialized in processing those visual stimuli that convey most social information, including bodies and especially faces.

In keeping with the principles of the temporal pathway, more anterior regions process increasingly more abstracted aspects of the face and are less dependent on the angle of viewing or the size of the stimulus on the retina. Many of these areas help establish whether a visual stimulus is a face and help identify the face. But although processing static information from faces such as the identity of the owner is a crucial skill in navigating the social world, equally important is the ability to extract more dynamic information from the face. Recent models of visual information processing suggest that the primate brain, at least in the case of simian primates, contains an extension of the temporal visual pathway running just above the traditional temporal pathway. This pathway is involved in processing more dynamic aspects of social stimuli.

The region in the middle part of the temporal cortex identified in the macaque social group size study is part of this expanded temporal visual pathway. It responds to faces but is not concerned with processing their identity. Rather, cells in this part of the brain respond to the location of eye gaze. Some cells fire if the eyes are gazing downward, others when the eyes are gazing to the left, and so on. Interestingly, if the eyes are not visible, the cells respond to the orientation of the head. The cells that previously responded when the eyes are gazing down now respond when the orientation of the head suggests that this is where the focus of the observed person's attention is. Likewise, if the head is not visible, the cells respond to the orientation indicated by the body. Thus, it seems that this area functions as a detector of the locus of attention of another person, using whatever is the best source of available information. This information and similar information obtained from body movements can then be used to calculate the meanings and intentions of the other person.

The human brain contains a very similar area. It receives information from the same regions and sends it to similar places in the brain. It is also located in the temporal lobe, although the expansion of this part of the brain in the human means the area has been pushed backward. This region has been studied from the first days that researchers were interested in the brain mechanisms of social information processing, because it seems involved in a classic social task: the false belief paradigm.

In this task, the participants are presented with a simple story, usually portraying two children. One child takes a ball and hides it somewhere in view of the other child. The first child then leaves the scene and the other child uncovers the ball and hides it somewhere else. The first child comes back. The question asked of the participant is: where will the first child look for the ball? Most adults will say that the child will look in the place they originally hid the ball, not in the place it actually is. This is correct, since the first child is missing some crucial knowledge, namely that the ball was moved while she was away. However, young children will often indicate the place where the ball actually is, rather than taking into account the knowledge of the first child. They cannot dissociate their knowledge from that of another person. This type of reasoning, dissociating the knowledge of another from your own, is considered a hallmark of human social information processing. It thus seems to rely on a region in the human brain very similar to the one that other primates use to process information about others' location of attention. Other primates might not quite process the information to the level we do, but the overall brain infrastructure seems very similar.

As this finding shows, complex social information processing does not come out of nothing, it is reusing the temporal visual pathway that was already present in the primate brain. This 'hijacking' of existing brain systems by another function can be seen in other social skills as well. One of the principles of temporal cortex organization that we talked about in the previous chapter was that of increasing abstraction. The farther forward you go in the temporal cortex, the more complete, observer-independent, and abstract the processed information becomes. Eventually, we saw, we find cells that can respond to learned categories.

In social life, we of course also categorize continuously. We might not like it, but we do tend to classify people into categories almost as soon as we meet them. Such heuristics help us navigate the large number of individuals we meet. If some of the temporal lobe functions have been hijacked by social information, could our placing of people into categories rely on the same system for learning categories we saw before?

My colleagues Suhas Vijayakumar and Egbert Hartstra decided to test this idea. They presented people with virtual avatars that belonged to different groups. The participants had to learn people's groups based on their clothing and then use knowledge of the groups to make inferences about people's preferences. Among the regions that showed increased activation when participants learned about group categories were regions in the temporal lobe, including a region in the most anterior part, just above the categorization area often observed in nonsocial studies. Studies from other groups confirm this result, that social knowledge is processed in regions of the temporal lobe just above those processing more worldly knowledge. Social information processing literally builds upon the temporal visual pathway that originally evolved to help simian primates forage better.

———

From sociality to perspective-taking

The temporal cortex can thus set the context of our social situation: who is present, what do they know, how are they feeling? But, as we saw in the previous chapter, the temporal cortex does not have direct access to the motor system. For that, we need the frontal cortex, which is able to integrate the sensory context with our goals and means. This allows us to deal with a complex and often dynamic environment. Those challenges are even more complex in a social context.

In a social context, we have to deal not only with the state of the world itself, but also with how somebody else perceives the world. For instance, I might know how something works, but my young child might not understand that. If I talk to her assuming she has the same knowledge I do, the conversation is quite short, she will get frustrated and concentrate on more interesting things, often involving the latest episode of *Peppa Pig*. If I do take her knowledge and perspective into account, the interaction will be much more fruitful and more pleasant for both of us. Children realize this very quickly, not only as receivers of information, but also as senders. My five-year-old daughter takes another tone when explaining something to me than when explaining something to her three-year-old brother. This turns out to be quite universal. Children of this age adjust their behavior to suit the knowledge of others, especially if they are often exposed to other children, such as in a daycare setting.

This kind of perspective-taking is difficult. For one thing, it can go on and on. Not only do we have to take into account the knowledge of one other person, sometimes we have to take into account what another person knows about the knowledge of another. Or even what another person thinks about what another person knows about the knowledge of another. This type of scenario is very prevalent in our social life. So much so that it forms the basis of most soap operas, situation comedy shows, and works of great literature. Although that does not mean we always consciously follow the whole train of thought of "he said she said he said . . ." it does mean that we have to be extremely flexible in our sense of perspective.

To investigate how our brain copes with this type of reasoning, the neuroscientists Giorgio Coricelli and Rosemarie Nagel presented volunteers with a classic social reasoning game. In this game, people must name a number between zero and 100. The person who names the number closest to two thirds of the average of all the numbers named by all participants wins. If you play this game and do not take social context into account at all, you would probably guess two thirds of the average value of all possible numbers, so ⅔ times 50 is 33. But you could think of your opponent doing that as well, so you should choose ⅔ of

33, which is 22. Or you think that's what your opponent would do, so you should choose ⅔ times 22 . . . etcetera.

A specific part of the frontal cortex helps us solve this type of game. It belongs to a family of regions related to the anterior cingulate cortex region we encountered in chapter 2 as an example of an association area that helped mammals deal with their cost-benefit analyses during foraging for high energy foods. That region, we argued, was capable of integrating information about the environment, the potential alternative foraging scenarios, and the current needs of the animal. It was an early example of the integration of context, goals, means, and outcomes. Theoretically, these are computations the frontal cortex needs to perform to make sense of the ever-changing foraging environment of simians and could be suitable to deal with the uncertain and unconstrained nature of social interactions. Could social information processing rely on similar structures in the frontal cortex, hijacking the system just as in the temporal cortex?

Nils Kolling, who designed the study of the role of the anterior cingulate cortex in foraging in chapter 2, Marius Braunsdorf, and I decided to test this idea. We designed a task to dissociate knowledge about the real world and knowledge about somebody else's idea about the world. In the constrained confines of an MRI scanner it is difficult to make the "world" particularly interesting. Our best attempt was to show the volunteers a screen consisting of different parts that could open one by one like little doors, a bit like an advent calendar. Behind each door was a cloud of blue and red dots. Over the course of different doors opening, the participants had to figure out whether there were more blue or more red dots. We know that the parietal stream—capable of keeping track of quantities—can extract this type of information from our visual input and pass it on to the frontal cortex. This gives the brain a good idea of the state of the world.

The crucial manipulation of our task was that another person, represented by an avatar on the screen, was also performing the task. But the avatar could not see all the open doors. Thus, our participant had an idea of the state of the world, and could develop an idea of the avatar's idea of the world by keeping track of which doors did not open for them. After each series of doors opening, we asked the participants what the state of the world was, where there were more blue or red dots. Then we asked them what they thought the avatar thought about the world. Depending on which doors opened for them, they could have the same view of the world, or their idea could differ.

It turns out our frontal cortex can represent these different types of information very well. A large part low down in the frontal cortex perfectly represented the state of the world, distinguishing whether there was lots of evidence in one direction or whether the numbers of blue and red dots were quite similar. But more important is what happened in the regions we identified before and that were also observed in the social reasoning game. These regions are closer to the output systems of the brain and need to be able to deal with the information relevant at a given time. This region also represented whether there was strong evidence in favor of red or blue. But there was a twist. When we asked the volunteers to tell us about the actual state of the world, this region represented that. But when we asked what the participant thought the avatar would say, the region adjusted its perspective; it now reflected what the avatar thought about the world.

Just as in the foraging domain, this part of the frontal cortex is capable of adjusting to the information relevant to the decision at hand. Probably this is what happens in my daughter's brain when she adjusts from talking to me to talking to her three-year-old brother. Or, perhaps, when she adjusts her perspective to my limited knowledge when explaining about Peppa Pig.

———

Recap: The fox and the dog

So where does that leave my experiment on my cat? Social life is not the standard way of living. In a world where other animals want to eat you and your conspecifics are competing for food and mates, being solitary for most of your life seems the safest option. In this situation, knowing if a pair of eyes is looking at you is a good strategy to keep safe. My cat can do this. But some species go further and exploit the potential advantages of foraging in groups.

Social life is complicated. Domestication shows both that not all animals can adapt to living in groups and how those that can adapt do it. Fight or flight responses that normally activate in the presence of others are dampened—tameness—and animals stay more receptive to the way other individuals view the world. It can turn wolves and foxes into dogs and doglike animals.

Social life is not a one size fits all. Primates need to survive in complicated social structures, and all seem to be able to detect where another is attending to use that information to make some inferences on what the other wants and knows. Humans have taken this to the next level, seemingly capable of following endless recursive steps through the social landscape and fully understanding the perspectives and beliefs of others and adapting their behavior accordingly. But what precisely drove the human brain to become so hypersocial?

6

The chimpanzee and the human

THE INTEGRATED, CULTURAL BRAIN

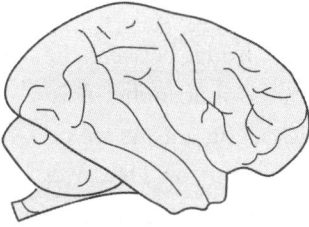

Chimpanzee

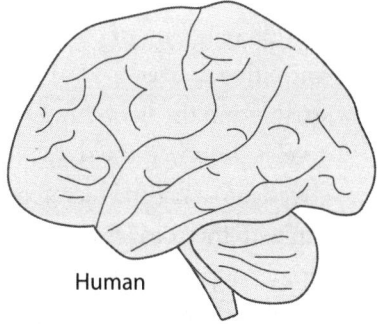

Human

There is something strange about humans. Evolutionarily speaking, we are just another ape. We are a branch of the great ape lineage that also contains orangutans, gorillas, bonobos, and chimpanzees. Just like other great apes we have big bodies, develop slowly, live long, and mostly give birth to only one offspring at a time. We also all lack a tail. But something about humans is different. Apes were the dominant primates 13 million years ago, during the Miocene. Apes lived in Asia, Africa, and even Europe. However, after the climate took a turn for the worse, great apes became less prominent and the monkeys took over as the dominant primates. That is still the situation today. Monkeys live in Asia, Africa, South and Middle America, and—in the case of the barbary macaque—Europe. Great apes, in contrast, are confined to the tropical forests of sub-Saharan Africa and Indonesia. Except one. Humans live permanently on every continent bar Antarctica. How did this happen?

Religion and science alike have long claimed that humans are unique among life forms. If not wholly distinct from the animal kingdom, then at least humans were thought to form in some way a favored branch of the tree of life. According to many of the world's religions, of course, humans are unique, favored by God and either created in His image or serving as a bridge between the earthly and the heavenly. Outside religion, the philosopher Aristoteles proposed the *scala naturae* in which all organisms are ordered into a natural hierarchy, with humans at the top. Descartes ran with this idea, arguing that apart from humans all animals are simply response machines, incapable of modifying their behavior through rational thought. In Darwin's time the biologist Richard Owen claimed that the human was the only great ape whose brain contained a specific structure, termed the *hippocampus minor*, which he believed endowed them with unique mental abilities.

Today, there is no shortage of scientific and not-so-scientific theories on what makes humans different from other animals. The explanations take an abundant variety of forms. Some explanations focus on features of our bodies, be it our strange form of locomotion on our hind limbs, our flexible hands capable of fine manipulations, our lack of body hair, our ability to sweat and blush, or our ability to run long distances. Explanations focusing on our brain emphasize its large size, its large frontal cortex, an abundance of dopamine molecules, an abundance of oxytocin molecules, among others.

Explanations focusing more on our cognitive abilities and our behavior emphasize our extreme sociality, our ability to use tools, our spoken language, our tendency to tell stories, our mastery of fire, our ability to count, our culture, our moral behavior, or our ability to create abstract concepts. There is a wide range of explanations. The search for a single explanation of what makes humans different from other species is so pervasive in the history of human thinking that Kensy Cooperrider of the Many Minds podcast coined the term "uniqual" to describe any aspect of humans that has been proposed as unique.

Over the last few decades, the pendulum of thinking about human uniqueness has swung the other way. Popular science writers never tire of telling us that our DNA is 98% similar to that of chimpanzees. It is now common to suggest that most of the behaviors most people would consider distinctive of humans are actually ape behaviors with a bit of human sauce. Or, the other way around, that nonhuman animals are capable of much more complex "humanlike" behavior than we originally thought. Perhaps most uniquals are not absolute, but more a matter of degree.

For example, most great apes can walk short distances on their hind limbs, but we are the only ape species for whom walking upright is the default mode of transport, rather than the occasional waggle. Chimpanzees use twigs as tools to fish termites out of hills, but that is a far cry from humans who use their smartphone as a more or less permanent extension of their body. Other animals also communicate, but our sophisticated system of language has not been found anywhere else in the animal kingdom.

What does that mean for the search for uniquals? Can we find anything that is a true uniqual or only ones that are relative? And if we do find anything, how did it come about? Can relative uniquals explain why our dominance of the planet is so overwhelming while our great ape relatives are at risk of going extinct? The problem with the search for uniquals is that we generally just compare behaviors or anatomical features between humans and another animal, without asking what they are for or how they came about. Take language. Yes, humans are the only species that has spoken language. But language itself requires cooperative individuals, active teaching, and nurturing of our young through a very long period of dependence. To understand language, we need to understand how and why these earlier necessities came about.

Similarly, if we want to understand how the human brain evolved, we have to look at what happened since the last ancestor of humans and nonhuman primates lived in the rainforests of Africa about 6.5 million years ago. Again, concentrating on the foraging challenges that our forebears encountered will be helpful. It is a journey that leads to our brain being much bigger and differently wired compared with even our closest animal cousins.

———

A quick who's who of human evolution

Humans are the children of Africa. Darwin speculated that humans originated in Africa based on the observation that most living great apes—our closest animal relatives—are found there. As we saw in the introduction, Dubois did not agree and searched for the remains of early humans in Asia. His success in finding *Homo erectus* notwithstanding, we now know that humanity did originate in Africa and dispersed across the rest of the world in a series of migrations.

One way we know this is from genetics—modern Africans have much greater genetic diversity than people living on the rest of the Earth, an indication that humans have been in Africa longer. Fossils of early humans are also found in Africa, most prominently in a long band running from Ethiopia through Uganda and Kenya, toward Tanzania. This part of East Africa now houses some of the great African lakes, such as Lake Turkana and Lake Victoria. This is the area of the East African Rift Valley. According to the British geographer Mark Maslin, climate change and changes in the environment in this part of the world set the stage for the evolution of humans.

About 20 million years ago, at the level of what is now the Democratic Republic of Congo, rainforest stretched across most of the east-west expanse of Africa. On both sides, this rainforest received a steady supply of moist air from the Atlantic and Indian Oceans. This is a tectonically active part of the world, though. A magmatic hot spot beneath the crust of the Earth started pushing part of East Africa upward. The crust expanded and stretched, creating a high-lying plateau.

Eventually—we are talking millions of years—the rocks started to fracture and faults appeared at the edges. These allowed the central part of the plateau to sink back down again, creating a valley with large rock formations on either side. These rock formations block the moist air coming in from the oceans, creating a rain shadow. As a result, the formerly lush rainforest area of the valley became drier. This effect was exaggerated by the formation of a second plateau, thousands of miles away in Asia. The formation of the Tibetan plateau caused a major change in the circulation of the air in the area, pulling away moister air from East Africa. By five million years ago, rainforest was no longer viable in the Rift Valley. Instead, a more fragmented landscape of different types of vegetation and alternating arid landscape and lakes took its place.

All this happened against a background of a changing climate. After a brief respite, the trend of cooling we saw in chapter 4 continued, but the climate also became increasingly unstable. The way in which the earth moves around the sun changes in a series of celestial cycles. The orbit around the sun is slightly elliptical, not perfectly round, and the ellipse stretches and shrinks. The tilt of the earth's axis also varies. As does the amount of wobble of the axis, like a spinning top does before it finally topples over. Together, these cycles influence how close the earth, or parts of the earth, get to the sun and thus how much energy different parts of the surface receive. As climate is essentially the redistribution of energy around the globe, warming a particular part more or less can have profound consequences.

The Rift Valley was one place where the consequences of such climate instability could cause profound changes in local habitats. These consequences became particularly strong after 2.7 million years ago. Around this time, North and South America joined, closing the Isthmus of Panama. This sent more moisture in the direction of the Arctic, creating a permanent ice cap. With this ice cap, the climate cycles have more cold "raw material" to play with. Depending on how the three cycles lined up, the northern ice moved southward, covering large parts of Eurasia and North America, or retreated. The period of advancing and retreating polar ice is what we commonly refer to as the Ice Age. It set the scene for the evolution of the human genus. And it is still ongoing, although for the last 11,000 years we have been enjoying an interglacial period with ice confined to the poles.

Against this background of a change from a rainforest across Africa to a more arid region under the influence of an increasingly volatile climate, the various descendants of the last common ape/human ancestor had to find their way. The ancestors of the lineage leading to chimpanzees and bonobos stayed in the rainforest. The ancestors of another lineage likely started out in the changing East African landscape and led, eventually, to us. Exactly how that lineage developed over time is difficult to reconstruct. A diverse group of species that lived in the period between the last common human-chimpanzee ancestor and modern humans have been identified, but it is impossible to determine exactly who begat whom. One species can be the ancestor of another or it can be a side branch on the evolutionary tree. Even more pressing, it is difficult to establish exactly how many species are identified. Some researchers— the "splitters"—tend to identify quite a few, while others—the "lumpers"—argue that observed differences between fossils are likely differences between individuals of the same species, rather than that they belong to different species. Still, it is possible to identify groups of proto-humans that most researchers will agree on. Let's introduce the cast.

The first group are the oldest known remains commonly thought to be on the line between the common ancestor and us. They consist of the 6–7 million year old *Sahelanthropus tchadensis* from Chad, the 6 million year old *Orrorin tugenensis* from Kenya, and a few species belonging to the genus *Ardipithecus* from between 5.8 and 4.3 million years ago. We really do not know much about any of these. As one paleoanthropologist pointed out, you could fit all the remains of these species together into a decent size shopping bag. From *Sahelanthropus* we have a nearly complete cranium and some teeth, jaw bones, and a few arm and leg bones. The situation for *Orrorin* is also quite dire: some loose teeth, parts of a jaw, and bits of arm, thigh, and finger bones. The situation is a bit better for *Ardipithecus*, with even a partial skeleton of a female of the species *Ardipithecus ramidus* recovered from the Rift Valley.

In general, all these species remind us very much of great apes, but some first signs of change are present. The *Ardipithecus* looks to be of a more slender, call it gracile, build. They look a bit more like a juvenile ape, which reminds some researchers of the changes we see in domesticated animals when they become more social. The shape of the pelvis of *Ardipithecus* and the thighbone of *Orrorin* allowed a more stable gait than the sideways and energetically costly waggle of chimpanzees when they walk on their hindlegs. The cranium of *Sahelanthropus* looks like the head was placed on top of the body with the spinal cord connecting from underneath, rather than at the back. Together, this suggests that these species were better adapted to walking upright on their back legs than nonhuman great apes are. Still, *Ardipithecus*'s long arms and big toe that was capable of grasping suggest that these creatures still spent a lot of time in the trees. It has even been argued that this is where bipedal walking started, as a quick way to move along branches while using the arms for balance, similar to what orangutans do.

One day, in what is now Tanzania, three proto-humans were walking along. One of them stepped in the footsteps of another, leaving superposed footprints. Volcanic ash from a nearby eruption rained down on their footprints, eventually cementing them. About 3.7 million years later, in 1976, Andrew Hill, an exhibition curator at the Yale Peabody Museum of Natural History, was visiting the paleontologist Mary Leakey at an excavation dig. Walking back to camp one evening, Andrew and his colleagues were playing around. He had to duck a ball of elephant dung thrown to him by a colleague. With his face close to the rock, he saw footprints fossilized by volcanic ash. He noticed antelopes, rhinos, and, in between, proto-humans— the footprints of our proto-humans millions of years earlier. The footprints are clearly from a bipedal creature. They are now commonly attributed to the next member of our cast, *Australo-pithecus*. Several almost complete skeletons of *Australopithecus* have been recovered, including a famous 40% skeleton of a female, nicknamed Lucy. (Convention in popular science books now dictates that I relay the story that she was named Lucy because the song "Lucy in the Sky with Diamonds" was played repeatedly at the party the evening the skeleton was discovered.) The australopithecines died out 1.8 million years ago, but they are generally considered to have begat the species that started the next genus in our story: *Homo*.

Humans belong to the genus *Homo*, in which we have assigned ourselves the not-quite-so-modest name *Homo sapiens*, the thinking man. Typically, the oldest species to be designated a member of the genus *Homo* is *Homo habilis*, the handy man. It was long thought that this was the first proto-human to manufacture and use tools and that this necessitated that they were smarter than their *Australopithecus* ancestors. The tools made by *H. habilis* were sharp-edged stone flakes that could be used for woodworking or cutting and cleaning hides. These tools earned *H. habilis* its name and position as the first member of the genus *Homo*. Now, however, we know that tools were already used by the australopithecines, which takes away some of *H. habilis*'s claim to fame. It is also increasingly recognized that many great apes use tools in the wild. Chimpanzees fish for termites with sticks and a gorilla has been observed to use a bit of tree trunk to build herself a bridge across a swampy area. Because great apes' tools are all made of wood, they do not fossilize easily. If early humans used the same tools as great apes, they might simply not have been preserved.

As a final nail in the coffin of *habilis*'s fame, researchers are now experimenting with the techniques used for manufacturing the stone tools of *habilis*. Although effective, *H. habilis*'s tools do not seem to require much foresight to be constructed. In other words, not too much intelligent planning went into making these stone axes. This idea was tested by a group of researchers at University College London. They asked three professional archeologists who had years of experience with the toolmaking techniques of early humans to construct either *H. habilis*-type tools or more advanced stone tools associated with later proto-humans, while their brain activity was probed using PET. As we might expect from our discussion of primate brain specializations in chapter 3, the parietal visual pathway was activated in the toolmaking tasks, but much more so in more modern toolmaking than in that of *H. habilis*. Perhaps the basic toolmaking did not require *H. habilis* to be as clever as was originally thought.

The more modern toolmaking tested by the researchers in London was the so-called Acheulean culture, from between 1.7 and 0.25 million years ago. The tools are much more complicated to construct, requiring foresight into their eventual construction. It likely took years of practice and active teaching for a person to master the technique. The tools were also much more diverse, including hand axes, picks, and likely spears. These tools are associated with a different proto-human, *Homo erectus*. If we were to award a prize for the most successful type of human, we would surely have to give it to *Homo erectus*, the species first discovered by Eugene Dubois in 1893. It probably first appeared around two million years ago, and some small population of subspecies might have survived until a mere 110,000 years ago. That gives it a species lifespan of almost two million years, 10 times as much as *H. sapiens* has managed so far.

As the name implies, *H. erectus* was a fully upright bipedal proto-human, with limb proportions, spine organization, and height all comparable to us. It also had a humanlike shoulder suitable for throwing objects—say, a spear—at high speed. The developmental trajectory of *H. erectus* was also likely much more similar to that of a human than an ape, with perhaps a similar growth spurt during their teens as we do. The difference in height between males and females was also much reduced, again similar to what we see in humans, indicating that perhaps their family structure was more like ours than that of an ape. As a major step for proto-humans, *H. erectus* is also the first proto-human whose remains can be found outside Africa, in Europe, India, China, and Indonesia.

H. erectus was succeeded in Africa by *Homo heidelbergensis*, which in turn made its way out of Africa and perhaps became the direct ancestor of a whole series of modern humans. In Europe these were the famous Neanderthals, in Asia the recently discovered Denisovans, and in Africa around 200,000 years ago the first *Homo sapiens*. *H. sapiens* dispersed out of Africa in multiple waves. When they met other species of humans that had gone before, they sometimes interbred, and always displaced them. Eventually, *H. sapiens* was the only human species left.

———

Early human foraging

As the ecology of East Africa changed, so did the foraging situation of the proto-humans. If we assume the last common ancestor of humans and chimpanzees was a forest-dwelling, large-bodied, fruit-eating ape, the reduction of rainforest when the climate cooled posed a problem for our ancestors. The ripe fruits become more dispersed and, probably, more seasonal as well.

One solution could be to retreat to your patch of forest and become an ultra-specialist. This is what the ancestors of the other surviving great apes probably did. Not so our first group of proto-human cast members. As we already saw, they became more bipedal. Bipedal locomotion is a much more efficient way to travel on the ground than the bent-over knuckle walking of some great apes, allowing proto-humans to travel greater distances on the ground between food patches. Evidence from fossilized teeth also points to a shift in their diet. Early proto-humans had bigger, thicker teeth and could chew more forcefully, which indicates that they likely consumed more tough, fibrous food. For modern great apes, this is fallback food when the preferred fruits are not available. For early humans, it became a more substantial part of their diet.

This continued in the second group, the australopithecines. Even more adapted to walking longer distances, they became ever more reliant on non-fruit. Likely they used their early tools to get to new parts of plants that their ancestors routinely ignored, such as the hidden roots, bulbs, or tubers. These need digging out, which takes an investment of time and effort, but is worth it if successful. Predicting the presence of these underground storage organs and making some members of a group available for this task would be a great foraging advantage.

The key evolutionary advantage of proto-humans was their adaptability. As we saw, the climate in East Africa became increasingly unreliable over time. Millennia of cold, dry climate when few resources were available could quickly change into millennia of wet climate when a lot of resources were available. This could then be followed by a cycle of highly variable climates, before a new cold and dry period started. In those climatological circumstances, specialists would find it hard to survive for long. In contrast, animals that are more adaptable, that can change their foraging strategy to suit the circumstances more flexibly, can thrive. This is what early proto-humans did. They developed the ability to learn quickly and deal with changes in their circumstances. This adaptability allowed them to exploit a greater variety of niches. Current and recent hunter-gatherer societies function this way as well, adapting their diet to the ecological circumstances to maximize their overall food return for time and effort invested.

Early *Homo* would become the first proper hunter-gatherers, inventing a foraging strategy that relied on gathering plant foods, including underground storage organs, supplemented with the occasional spoils of hunt. This strategy required something new, something that no primate had managed before. It required close cooperation by a large group of individuals.

We saw in chapter 5 that primate species are generally quite social. They interact with large groups of individuals and have complicated hierarchy structures. They know about their own relationship to others, but also the relationships of others with others. This ability proved crucial to the early humans. To be able to move to a hunter-gatherer strategy they needed to be able to divide up tasks. Some people gather and provide a reliable stream of incoming resources. Others hunt for the occasional big prize. This can only work if the gatherers will return home with their spoils and share with the hunters if they were unsuccessful and, in return, the hunters share their spoils— which are probably too big anyway to consume by themselves— when they are successful. This requires long-term cooperation even with individuals with whom you have no genetic relationship. But it is a successful strategy that will allow every individual in the group access to more resources than each of them would have if they had to fend for themselves. In other words, a prime strategy for proto-humans to deal with the foraging problems associated with the varying climate was to become a cooperative social species.

Being a cooperative species requires more than sharing the spoils. Successful hunting requires coordination of the behavior of individuals. It requires an understanding of having a common goal and coordinating one's behavior toward this goal. This requires me to understand what it is that you know, so I can help you toward our common goal. Psychologists call this ability "theory of mind," our ability to have an appreciation of the content of the mind of others—what they know, feel, want, and believe.

As we saw in the previous chapter, this level of mentalizing is beyond the abilities of most other primates, who understand the actions of others only in terms of the competition for resources. Our collaborative foraging behavior also requires that we can detect and punish freeloaders. If all the spoils are shared, then it is beneficial for you as an individual to occasionally goof off and just enjoy the rewards of others' hard work. But this would be to the detriment of the group. Humans are very good at detecting free riders and punishing them. We judge people who do not play their part. So much so, that humans usually work very hard to gain a good social reputation, since a bad reputation might lead to exclusion from the group and hunger. In fact, it could be argued that this is the basis of human morality. We have not just become fairer to others, we enforce strict social norms in our group that become part of how we define our identity.

Our social abilities started to allow us to pass on knowledge. Not only are we inclined to help others and coordinate our actions with them, which other primates cannot, we also understand how to instruct others through our theory of mind. We use our social abilities to actively instruct our young. Chimpanzees do not do this. In a way, chimpanzee children have to reinvent the wheel again every generation. It might be a little bit easier for them than for their parents, since they tend to hang around their parents and play with the things they leave around. If all the bits and pieces of wheel are available it might be easier to reinvent one, but that is nothing compared to the explicit instructions our young get. This allowed a reservoir of knowledge to be built up—a handy tool if your climate can turn and you might have to rely on knowledge no person alive has ever had to use. It also allows our species to build on the inventions of the previous generations, creating not only a reservoir of knowledge, but also a cumulative body of knowledge.

The ultimate example of our collaborative strategy is of course our communication through language. We do not know when language appeared. Some estimates put it quite recently, about 70,000 years ago. This would put it long after the first appearance of anatomically modern humans, 200,000 years ago. Others argue that the complexity of objects recently found in the archaeological record point to sophisticated information exchange already at the earliest appearance of modern humans. We may never know. But what we do understand now, is that language is of no use to you unless you are already a collaborative species.

Language must have begun as a tool for coordination and cooperation. Sure, a competitive species might use language to threaten competitors more effectively and generate a nice repertoire of insults, but that hardly justifies the massive evolutionary investment the development of language must have required. To create language, you already have to want to communicate. Think of our young. As soon as they require any form of speech, they spend all their time communicating whatever comes into their minds to their caregivers. They are wired up to share information. Only a commitment to a communal existence makes it worthwhile to invest in language. With that language, coordination of tasks, collective problem solving, and passing knowledge across generations becomes much easier. It might very well have been the step that allows our current ecological dominance. The story of Babel, of humans who all speak the same language becoming too powerful, has an important truth in it.

———

Comparing human and nonhuman brains

That our behavior changed over the course of evolution is clear. But how did our brain change to support these behaviors? The acknowledged fact about the human brain is that it is big. In terms of body size, we are not that different from chimpanzees, but our brain is three times as large. This was not a gradual increase since the last common ancestor. Lucy the *Australopithecus* had a chimp-like brain.

For a long time after the split of the chimpanzee and human lineages, brain size did not seem to increase drastically. Based on the space inside the skull, we can estimate that our first and second groups of cast members, up to *Australopithecus*, did not exceed 450 cubic centimeters—about the range of chimpanzees and gorillas. This was despite their bipedalism. Apparently, walking upright was not what required a big brain. *Homo habilis*, our erstwhile first toolmaker, has a slightly bigger brain at 600 cc. Still not a dramatic increase. As we already saw, early stone tools also do not require massive brain power above that of a chimpanzee. Dramatic expansion of brain size only starts to occur with the next members of the *Homo* group of our cast.

Homo erectus's long life span as a species saw quite an increase in brain size, from 960–980 cc early on to 1250 cc. This last value overlaps with the range we see in humans, who average around 1400 cc. What's more, the relative size of the brain compared with the body increased in *H. erectus*, strongly suggesting that brain power was selected for in this species. Thus, since the time of our common ancestor with chimpanzees, it seems our brain stayed relatively similar in size until about two million years ago. Since then, it has expanded to three times that of the chimpanzee.

But does our larger brain also pack more computational power? In other words, does it also contain three times as many neurons as the chimpanzee brain? For this, we need to return to Suzana Herculano-Houzel's brain soup. Remember, she found a way to effectively wash out everything but the cell nucleus and then stain the nuclei belonging to neurons. This gave her a way to count how many neurons a whole brain contains. She showed that compared with other mammals, primate brains are really efficient at packaging a lot of neurons close together.

For the human brain, it turns out the total number of neurons is about 86 billion. That is a lot, but is it exceptional? If you compare the size of the brain and the number of neurons across primates, the human brain falls exactly where you could predict it. In fact, our brain contains pretty much the number of neurons we would expect for a generic primate of our size; however, our brain is very big in terms of volume for a great ape, rather than compared with all primates. It turns out that great apes actually have fewer neurons for their body size than monkeys. This makes for a nice compromise between the "humans are unique" and "humans are just expanded apes" crowds. The human brain is extremely big with extremely many neurons, but it is what you would predict based on what we know about other primate brains. As Herculano-Houzel summarizes her finding: in terms of numbers of neurons, our brain is remarkable, but not exceptional.

Does that mean our brain is really a scaled-up generic primate brain? In other words, do we have the same number of regions and the same connections between regions, just bigger? Or did something change in the way our brain is organized? For a long time, this was difficult to test. Mapping out brain regions and the connections between regions was painstaking work, requiring sacrificing many animals and looking at section after section of brain tissue through the microscope. By identifying the thickness of the different layers of the neocortex and recording the types of neurons, it was possible to map out different regions.

Mapping out a single species' brain took years. Understandably, this work was done for only a few species. That animals needed to be sacrificed for this work also means that we know very little about the species most related to us, as it is considered both unethical and impractical to undertake such a project. But just as was the case for learning about brain activity, again it is neuroimaging techniques such as MRI that now enable new research. Animals can be scanned in the MRI scanner, just as humans can. This opens up a whole new array of possibilities.

In the early 2000s, the neuroscientist David van Essen realized that this is the perfect way to test the idea that the human brain is a blown-up version of a smaller monkey brain. He took a scan of a macaque monkey brain and one of a human brain. He then used a computer algorithm to enlarge the monkey brain scan to the size of the human brain. That way, van Essen could test whether all the regions needed to be enlarged by the same amount to get from the monkey to the human brain. This was not the case. When van Essen blew up the monkey brain, he found that the upsized monkey association cortex was nowhere near the size of the human's. Remember that the neocortex consists of primary areas that have direct connections with information from the outside, for instance receiving visual information or directing muscle movements, and association regions that mostly further process information within the neocortex. The experiment made it clear that human brain is not simply a scaled-up monkey brain; different parts of the neocortex scale differently. Across different primate brains, the larger the brain gets, the more of the brain is occupied by the association cortex. The human brain, in other words, devotes much more space to deeper processing of information.

My own adventures in comparing the brains of different primate species started in 2010. When I arrived at Oxford for my postdoc, I joined Matthew Rushworth's lab that had a unique combination of skills. It performed functional neuroimaging experiments in humans to find out how the human brain makes decisions. Nils Kolling's experiments that I discuss in chapter 2 were part of that line of research. But the lab also worked with macaque monkeys. Most neuroscience labs do one or the other. It is hard enough becoming an expert in the study of one brain and all the associated techniques, let alone two. By combining the two, the lab was in a unique position. It was able to directly compare results in the two species. A fantastic opportunity.

Setting up MRI scanning of macaques was not an easy task. Although the lab had received a big grant to do this work and a new MRI scanner had been installed, the project was still in the early stage. Together with a team of young colleagues, we made this our project. We got together monkey researchers, veterinarians to ensure the monkeys' health would be assured at all times, MRI physicists to make sure we could get good signals from the MRI scanner, data analysis experts to process the images; it was quite a team.

My main responsibility in those early days, apart from helping to analyze the data, was running the scanner itself. That involved pressing the right buttons to acquire the images, but also a fair share of guarding the highly magnetic scanner environment to ensure that nobody who was not properly de-metaled would enter. Quite a task when so many people were involved. All procedures of course were first approved by the appropriate authorities to ensure that we maintained the very highest standards in terms of ethics and monkey and human safety.

We scanned macaque monkeys using a functional MRI protocol, the same as you would to measure task-related activation in the human brain. However, rather than giving the monkey a task, we lightly anesthetized them. Such anesthetics do not shut down the brain, there still is a lot of spontaneous activity going on, just as when we sleep. From earlier research in humans, it was known that regions that have strong connections with one another also show a similar pattern of spontaneous brain activation. By comparing the spontaneous activation of different parts of the brain, we could thus get an indirect measure of whether they were connected to each other.

Studying these patterns of spontaneous activation allowed us to define each region of the brain in terms of its connections to the rest of the brain. Such a pattern of connections is unique for each region, it is almost a "fingerprint" for that region. We could then do the same for the human brain and perform a little detective work. We would obtain a connectivity fingerprint of a human brain region and then search the whole of the monkey brain to see if we could find the same fingerprint. If we did, we knew that the region was present in both brains. If not, we had a candidate unique brain region.

We set out to systematically compare the association cortex of the human and the macaque monkey. We concentrated in particular on the frontal cortex. As we saw in chapter 4, this part is especially enlarged in simian primates and is often identified as a part of the brain most developed in humans. If there was anything special about the human brain, surely this is where it would be. Studies like this are always hard to communicate to other researchers, because they will always disappoint somebody. If the researcher is interested in using the monkey as a "model species" for the human brain, perhaps using it to understand something about our brain that cannot be studied directly in the human, they hope that we are mostly big monkeys. Finding little or no differences between human and monkey brains is a calming result for this line of research. Most people are more interested in differences—they want to find something that might explain our human abilities. Findings of uniquals tend to be more sensational.

In this case, most of our findings showed that most brain regions found in human brains look similar to regions in the macaque monkey. Even parts of the brain that are thought to help in the processing of language had counterparts in the monkey; however, we did find some differences. One in particular caught our attention. In the most anterior part of the prefrontal cortex, called the frontal pole, we found a region that looked distinct from anything we could find in the macaque. Its pattern of connections was similar to neighboring regions, but not quite the same. It had stronger connections to parts of the parietal cortex that David van Essen's blowing-up-the-macaque-brain-to-the-size-of-a-human-brain study had shown were disproportionately expanded in the human brain. So, the region was slightly different from other prefrontal regions, it did not have a clear homologue in the macaque, and it was connected to regions in other parts of the cortex that had also expanded disproportionally in the human brain. Had we found our first uniqual?

At the time, the frontal pole was the focus of quite a lot of research in different labs across the world. In the lab where I worked, people had shown that when humans make complex decisions where they have to keep in mind many different alternatives of action, this region is very important in keeping information online, even if it is not used at the time but will be useful later. The French neuroscientist Etienne Koechlin termed this holding in mind of information during other work "cognitive branching." He suggested this capability forms the basis of many of the high-level problem solving and planning abilities often associated with human reasoning. In other words, he suggested the frontal pole indeed might be the basis of something uniquely human.

On the other side of the world, the California-based neuroscientist Silvia Bunge was performing a series of experiments on the neural basis of another behavior that is particularly well developed in the human: relational reasoning. She asked people to complete sequences such as "snow is to snowflake as army is to . . ." (soldier). Or people were told a series of relationships, such as a green ball is heavier than an orange one, purple is heavier than green, and blue is equally heavy as red. They then had to infer which is heavier, purple or orange? (purple) All these tasks showed activation of the frontal pole. These behaviors are very similar to the types of reasoning, such as fast categorization, we discussed in chapter 4 as relying on simian brain innovations, in particular the prefrontal cortex. And Silva Bunge found that they all consistently activated a region in the most anterior part of the frontal pole. She suggested that these behaviors are distinctly human and that this would fit very well with their relying on a region found only in the human brain.

One of the first experiments I did when I started a research group of my own was to test the idea that the tasks that Silvia Bunge used indeed activated the region we had identified in our macaque/human comparisons. I asked participants in the fMRI scanner to perform variants of the tasks that Silvia Bunge had developed and then used my own techniques to identify the frontal pole region based on their brain anatomy. I found a good overlap—the human relational region tasks indeed activated our frontal pole region. Here we had some first evidence that we could use MRI to establish differences between the brains of different primate species and that these differences relate to some behaviors that are particularly well developed in humans.

The behavioral conditions the frontal pole seems to support also fit with our notion that humans deal with their volatile environment by integrating more information, learning more quickly, and using mental trial and error. All of these fit with what we knew about this region: it is larger in the human brain, it is better connected with other parts of the brain, it matures late in development, and it shows the activations we would expect.

Of course, to truly claim this is a region unique to the human brain, we need to test whether it exists in primates more closely related to us, such as chimpanzees. We also should never link a complex behavior to just one brain region, as brain regions never work in isolation. And, sure enough, the theme of brain organization across species was definitely similar, following the primate patterns we saw in chapters 3 and 4. Still, we had shown there might be more to our brain than scaling up the monkey brain. The next step was to see if we could expand the research line. How far could we push MRI as a comparative tool?

———

Connecting the dots

In 2007, Jim Rilling arrived at the University of Oxford's MRI center to work on a collaborative research project. He brought a scientific treasure with him to explore. Together with his colleagues at the Yerkes Primate Center in Atlanta, he had acquired some of the first MRI scans from chimpanzees. Although chimpanzees are, together with bonobos, our closest animal relative, we knew almost nothing about how their brains are organized. We knew a lot about the macaque monkey, as it has been the subject of scientific investigation for decades. But the big, long-living, and closely related chimpanzee was never considered suitable for neuroscientific research, especially when these methods invariably involved the death of the animal. But now, using MRI, these wonderful animals could be studied without any adverse consequences to themselves.

Rilling and his colleagues did not collect brain activity in their scans but concentrated on a measure of brain structure, of anatomy. MRI can be used to collect many different types of information from the same brain. We can look at its functional activation as Nils Kolling did in his foraging experiments in chapter 2, spontaneous activity as we did in our early macaque studies described above, and structural scans to look at the shape and folding of a brain.

In the early 2000s, people had started collecting an even different type of signal. They started looking at the movement of water molecules in the brain. In certain parts of the tissue, water molecules can move around relatively unconstrained. But some types of tissue constrain its movement. One of those is myelin, the fatty tissue that insulates the white matter fibers that form the information highways in our brain. Water cannot penetrate the myelin, so it has to travel along the white matter fiber. Therefore, by using MRI to quantify how water moves through the brain, we can visualize these highways. This, in turn, can tell us a lot about how any specific brain is organized. Since it is a measure of anatomy, the data can even be acquired post-mortem, after the animals have died of natural causes; this is what Rilling did.

The study by Rilling concentrated on one particular white matter connection of the brain. It had long been known that there is a prominent set of fibers that connect the back of the temporal cortex with the lower part of the frontal cortex and that this set of fibers was important for the processing of spoken language. In the second half of the nineteenth century, the German physician Carl Wernicke presented a model of language processing in the brain, based mostly on lesions he observed in patients. Like many at the time, he saw the brain as a mosaic of areas containing different types of "memory images." The temporal lobe, he proposed, contains "auditory images" that are activated when somebody hears speech. The frontal lobe, in contrast, was thought to contain "motor images" that are important when we want to initiate movement, including speech. If we want to speak successfully, which includes monitoring what we say, information needs to be exchanged between these brain areas. This, he thought, is done via a prominent arc-shaped white matter pathway connecting the temporal and frontal cortexes, the arcuate fascicle. Damage anywhere in this arcuate pathway would lead to disorders of speech, or aphasia.

Since the nineteenth century, our models of language processing have grown more sophisticated, but the importance of the arcuate fascicle is still a prominent feature of most models. Rilling's study set out to determine whether this pathway, so important for a unique human behavior, is different in the human brain, compared with that of the chimpanzee. The results looked clearcut. The human arcuate fascicle extends much further into the temporal lobe than that of the macaque monkey or even the chimpanzee. Apparently, our frontal and temporal cortexes have a lot more to say to one another than those of other species. This finding of the expanded arcuate has become an almost iconic example of a major difference between the human and nonhuman primate brain. It has stood the test of time.

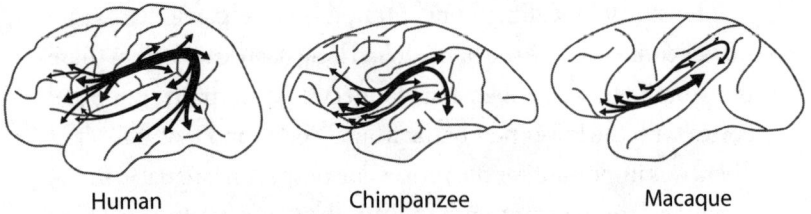

Human   Chimpanzee   Macaque

FIGURE 6.1. The arcuate fascicle is a white matter bundle that is far more extended in the human brain than in the brain of chimpanzees and macaque monkeys. It is the top bundle in the figure, connecting the frontal and temporal cortexes in the human brain. This is likely one of the specializations in our brain underlying language. Modified from figure 2b of Rilling JK, Glasser MF, Preuss TM, Ma Z, Zhao T, Hu X, Behrens TEJ (2008) The evolution of the arcuate fasciculus revealed with comparative DTI. *Nature Neuroscience* 11:426–428.

Such a big claim is always going to be controversial, but it has now been replicated by a number of other research groups, even using different methods. My own group has built on it as well. We were interested in what drove this expansion of the connection. In our view, two things could have happened. First, we already knew from the macaque-to-human-brain-expansion study by David van Essen that the human brain has a lot more frontal and temporal lobe than we would expect for a standard primate. If one of the areas connected by the arcuate has expanded, the arcuate would of course look much bigger. Thus, the arcuate expansion could simply be a result of more brain power. The alternative hypothesis is that the arcuate in the human is really connecting areas that are not connected in other brains. In other words, rather than "just" connecting expanded areas, it allows the human brain to do something fundamentally new. This more exciting hypothesis was the one suggested in the original study by Jim Rilling.

We developed a way to disentangle these various hypotheses. We tried to model what would happen to a macaque or chimpanzee brain if it enlarged and reorganized to a human brain and whether this would account for how the actual white matter tracts looked. This worked really well for a number of tracts, but not for the arcuate. This tract connected to completely new areas in the human, compared with the macaque or even the chimpanzee. The human brain really is wired up differently.

The Rilling study not only showed interesting differences between the human and chimpanzee brains, it also provided proof that the technique of using MRI to visualize the connections of the brain could be used successfully to compare brains of different species. This opened up a lot of possibilities. In theory, we could expand on the example of Rilling and collect brains of animals that had died of natural causes in zoos and learn about all these different brains, without having to harm any animal. This would be a large undertaking, involving veterinarians at zoos to help collect the brains after an animal dies, MRI physics experts to optimize the data collection, data scientists to develop new techniques to analyze the data, and neuroscientists to figure out what it all means.

My collaborators and I decided to take the plunge. Studying a large range of primates and establishing how their brains differ would give us an opportunity to understand the human brain within the context of other primates. This work is ongoing and we and other labs are slowly gaining insights into how primate brains—and ours in particular—differ. But some common themes are emerging. As the human brain is a primate brain, it is easiest to describe in terms of the common primate we already discussed: a visually oriented brain with different pathways processing different aspects of the world, including a parietal pathway for visually mediating actions originally due to the original arboreal niche, a temporal pathway setting the foraging context, and a frontal cortex expediting information processing to help an animal achieve its goals. These trends all continue in the human brain.

As we saw above, the human frontal cortex expanded and possibly contains new regions. These regions further expand the capacity of the frontal cortex to further process information, re-representing it in more abstract ways and combining more information. The parietal pathway also expanded, as we saw in chapter 3. The system originally used to compute how we transform information from the 2-D representation of the environment on our eye's retina to the guidance of the various muscles needed to effect change in the environment got co-opted for a much wider range of tasks. Representation and manipulation of information in different coordinate frames is very important for what we tend to call "general intelligence." The expansion of the parietal pathway and a dramatic increase in the information exchange between parietal and frontal cortexes facilitated this behavior.

But the most surprising differences we found were elsewhere in the brain. In one project, we tried to compare the organization of the entire macaque monkey brain with that of the human to see which areas showed the greatest between-species differences. We created a map that color coded each area of the human brain. The lighter the color of a region, the more likely that region could also be found in the macaque brain. The darker a region, the more different it is from any region in the macaque brain.

As expected, the primary visual cortex was light in color, just like the primary motor cortex. There were some darker areas in the areas we described above, in the frontal cortex and parts of the parietal visual stream. But what jumped out of the image was a big dark blob in the vicinity of the temporal visual stream. The main hot spot of differences was there, in the middle of the

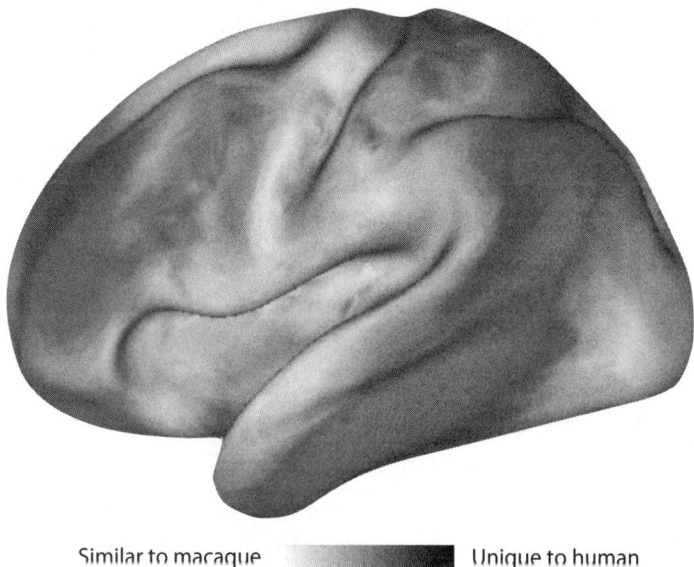

Similar to macaque   Unique to human

FIGURE 6.2. Map of the human neocortex, color-coded to indicate whether similar areas exist in the macaque monkey (light) or are whether they are unique to the human brain (dark).

temporal cortex. We already saw in earlier chapters that the temporal visual stream expanded in monkeys and apes, the simian primates. Now, intrigued by our map, we decided to zoom in on this area a bit further in our database of different primate brain scans.

We found that the temporal visual pathway dramatically expanded in apes, compared with monkeys; it does not form a single pathway, but a series of parallel streams. This trend continued in the human lineage, with existing primate trends present, but often in a modified form. For instance, areas processing social information expanded and diversified. As we saw in chapter 5, we looked at an area in the ventral stream that in the macaque monkey processes the direction of gaze of another individual. This is handy information to have if you live with conspecifics; knowing where they are looking informs you of their desires and intentions. In the human brain, a series of similar areas exist. Some still code information about eye gaze, but some have expanded their function to make inferences about the mental state of others. Our ability to infer the mind state of others—essential for our cooperative solution to our foraging challenges—is built on the more general primate ability for processing information about others.

A similar theme, a primate brain organization with modifications, is true for the increasing abstraction of information processing along the temporal visual pathway. Information from the different senses is much more integrated in our brain. In turn, this information is communicated much more widely throughout the brain. The integration of sensory context from the temporal visual stream with the goal information in the frontal cortex is much more prominent in the human brain compared with all the other primates we tested. The parietal and temporal visual streams themselves are also more interconnected, allowing much more direct and earlier information exchange.

Together, these changes allow information to be represented in a much more abstract way and be combined to a much higher level. We do not represent just a picture of a hammer in our brain, we associate it with a particular sound, a particular movement, a type of goal. We can even come up with very abstract concepts like companies and communicate these to one another. In our brain, we represent the "meaning" of something in a very different way from other primates.

———

Recap: The chimpanzee and the human

Where does that leave our search for the uniqual? Does our brain have human specializations that stand out, or are we just another primate? The answer, of course, is both. Our brain follows the primate template, which in turn is a variant of the vertebrate template. But, just like any brain, ours contains some modifications to the theme, and they have had a dramatic impact on our abilities as a species.

Take again the example of the temporal cortex. If we try to summarize how it is organized in primates, we commonly describe the visual pathway with its increasingly abstract representation of information the farther forward we move. While most of this part of the brain is devoted to processing visual information coming in from the occipital cortex at the back of the brain, we also commonly note that there is auditory information coming into the top of the temporal cortex.

Both principles are true in the human brain. But the incoming expanded arcuate fascicle that Rilling identified provides a whole extra dimension. Suddenly the middle and back of the temporal cortex are plugged directly into the parietal visual stream and the frontal cortex. As a result, our information processing is much more integrated. Concepts mean different things. Language, our most human behavior, becomes possible. But it can be understood only in the context of the primate template. Our evolutionary history cannot be ignored. In a sense, our brain is just another retrofitted primate brain.

7

The bird, the rat, and the human

RETROFITTING THE BRAIN

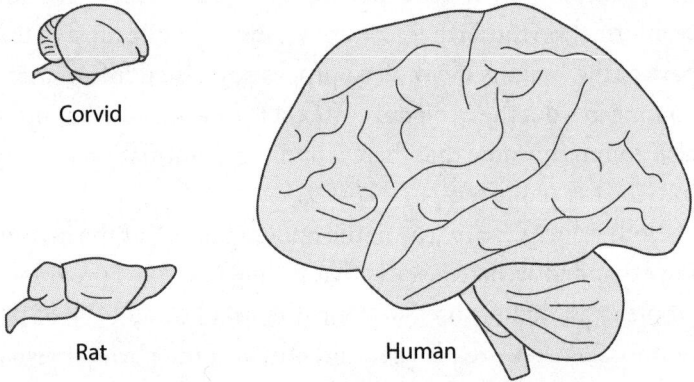

Corvid

Rat

Human

About two billion years ago, a small single-celled organism got swallowed up by another single-celled organism. It seemed an inconsequential event. Things like this happen all the time. But the effects could not have been more dramatic. The absorbed cell was a bacterium capable of producing hydrogen. The absorber, for want of a better term, was an archaeon. The Archaea are single celled organisms, like bacteria, but radically different in their internal chemistry. This archaeon needed two gases to grow: hydrogen and carbon dioxide. Absorbing the hydrogen-producing bacterium solved one of its problems. It turned out to be much more efficient for the archaeon to forage for food for its bacterium than to search for hydrogen itself. A new way of life was born through this symbiosis.

Now, the reminiscence of that bacterium is in our mitochondria, the cell organelles that produce most of the energy required to power our cells' biochemistry. Its impact on our lives is profound. Being able to produce energy much more efficiently opened the way to eukaryotic cells and therefore to complex, multicellular life on our planet. Without multicellular life, there would be no complex animals. Without animals, there would be no brains.

Events like this make you wonder. What would happen if we could set evolution running again? Would we end up with the same type of life we see around us? We saw in the previous chapters how brains have become diverse over the course of evolution. Early vertebrates came up with a brain that has all the major components we still use today. Some vertebrates moved from the sea to the land, adapting their brain in the process, and begat the amniotes. Amniotes begat reptiles and mammals. Mammals invested in a six-layered neocortex. Different mammals' brains adapted to their particular niches in various ways. Primates adapted to an arboreal niche of a visually guided grasper, developing parallel cortical pathways for processing visual information. Simian primates elaborated upon these pathways, and humans in particular found ways to integrate the information across these systems. As a result of all these splits of different lineages in the evolutionary tree, each adapting the brain to their needs, we now have a fantastic diversity of brains around us. But is the nature of brain evolution such that everything is stitched together by chance events, or is there some regularity that means certain outcomes were inevitable? In other words, are there some constant laws of evolution, or are we merely dealing with a chaotic process?

On the one hand, influential one-off events that determine the course of life seem abundant. The absorption that led to mitochondria and enabled the evolution of complex life is one example. The asteroid that triggered the dinosaur extinction and ushered in the age of mammals is another freak event that had dramatic consequences. Without it, mammals might still only be small nocturnal animals living in a reptile-dominated world. All seem to suggest an endless succession of chance events. If we could set evolution running again, the possible outcomes seem virtually endless.

On the other hand, the possibilities are not infinite. All life needs to obey the laws of physics. On any planet, gravity dictates that land animals need to adapt their bodies in a different way from animals who live in fluid. Similarly, if larger animals evolve that need to communicate with one another, the number of channels they have to do so is limited and dependent on where they live their lives. This is why dolphins do not communicate through large colored feathers and birds do not use odors to mark their territories. Variation might theoretically be limitless, but the point of natural selection is that variations that fulfill a certain function remain—and those functions are best performed under a limited set of circumstances.

Another major constraint on evolution is that we cannot simply modify a working system ad libitum. Take whales as an example. I write this paragraph sitting in the coffee area of Oxford's Natural History Museum. Suspended from the ceiling are the skeletons of various species of whales. In quite a few of them, you can see tiny bones at the back of the body. They are vestigial bones related to the hind legs of their ancestors. Whales descend from four-legged mammals and, while their front legs were converted to fins, the hind legs served no purpose. But evolution couldn't quite get rid of them.

The complexity of the genome, with comparatively few genes having to provide a way to steer the development of an extremely complex body, means that evolution just cannot simply add or remove bits. It works mostly by modifying the existing plan, and that, of course, means tinkering with a complex system without destroying the bits that keep it working. This is why the great explosion in variation we see early in vertebrate evolution was preceded by a duplication of the genome. It allowed evolution to tinker with one copy of the genome, modifying genes to achieve new functions, while the other copy would keep working what already worked. The room for evolution to change stuff is very limited.

What does that mean for brain evolution? Are the brains we observe just the luck of the draw in an infinite space of possibilities, or are the options much more limited? The answer, as it is most of the time in science, is that it's a little of both. Evolution thrives on flexibility, and over the last billions of years, through continental drifts, climate changes, and catastrophes, it has had ample chances to experiment and come up with a pretty astonishing array of innovations. But the constraints that work on any evolutionary process also apply to the brain.

Any brain must follow the laws of physics, which limits flexibility in brain design—no bird will have a five kilogram brain to fly around with, and our children have to be born with a relatively immature brain to allow passage through the birth canal.

Brains are also expensive to run, which again means there are limits on how big they can be; acquiring enough energy to run them is a major foraging challenge in itself. And, crucially, brain evolution is also very constrained by a species' own history. Once an animal has committed to a certain brain design, it cannot simply abandon it without disastrous consequences. Thus, the search for flexibility is a search for modifications within the confines of previous solutions. Brain evolution is not the engineer's dream of building a Ferrari racing car from scratch, it is the hobbyist's retrofitting and modifying the old Buick in the garage by adding a louder exhaust and painting some stripes on the doors.

Sometimes this retrofitting and upgrading leads to strange but inoffensive solutions. In the human brain, we have a strange white matter connection called the fornix. It connects the hippocampus with nearby parts of the forebrain. Even though these structures are located next to each other, the fornix follows a strange route, running from the front of the hippocampus all the way along its length toward the back of the brain before curving upward around the thalamus and then back to connect to the parts of the forebrain located just in front of the hippocampus. To an engineer, this makes no sense. But it does in the context of brain evolution. When parts of the brain expanded, different structures moved around. The hippocampus is one of them—it rotated around the thalamus when the neocortex expanded and dragged the fornix along with it. In the mouse, where the neocortex is much smaller, the fornix looks much more sensible, a short fiber bundle connecting neighboring areas. In the primate, it makes a strange, long detour, because evolution could not simply unplug the connection and put it in a different place, as we would do when reconnecting new bits of our car stereo.

This chapter is about such limitations in the ways evolution can modify complex brains. It is about creating new variation while dealing with the constraints of physical laws and prior constructions. In other words, it is about retrofitting the brain. There are three types of retrofitting we need to understand. The first is what happens when the same solution to a problem is found in different evolutionary lineages—in other words, if two animal groups with divergent brains need to address the same problem. In the car example, if I want my engine to provide more power, I would come up with a different way of doing that for an old Ford Galaxy diesel engine than for a Tesla fully electric engine. The results would hopefully be fairly similar—I would be able to drive faster in either car—but the ways to get there would be different. In evolution this is called *convergence*.

The second case we need to understand is when an existing structure is adapted for a new purpose. It was nice that cars had a display on the dashboard that indicated which compact disc you were playing, but when compact discs went the way of the dinosaur, the display could usefully be converted to house the satnav or the display for a rearview camera. An existing structure is still there performing its old function—displaying information to the driver—but using new information. In a sense, we are talking about *upgrading* an existing structure.

The third case is similar in some respects. We use an old structure, but now convert it for a new function over the course of our lifetime. In the car metaphor, I can convert the trunk of my old convertible to house the speakers of my new sound system. The trunk had an original purpose, but my cultural circumstances have made me reappropriate it for a new function, that of sound signaling—presumably to advertise the car owner's wonderful taste in music to potential partners in the neighborhood. Something similar has often happened in brain evolution, including in our human brain, and it has been termed *neural recycling*.

———

Small brains

A forager's life is hard. Periods of food abundance are just as likely as long periods when it's slim pickings. During times of plenty, smart foragers store food away for leaner times. Jay is one of them. He has a number of places where he stores his food. Hide is a better word. There are thieves around that will empty your stash the moment you leave it unguarded. Jay should know, he has emptied others' stashes often enough when the need was high. When Jay feels he is observed he will stash his food, but he quickly restashes it somewhere else as soon as the uninvited observer is gone. He has an extraordinary memory and can remember his many stashes months later with very high accuracy. He is also a careful planner, anticipating which foods he will need in a couple of months' time, rather than storing what his hunger now would suggest. This arsenal of skills has made him a survivor while others perished.

You probably guessed that Jay is not human. He is a scrub jay, a tiny bird from the Corvid family. To survive during periods of low food availability, scrub jays cache food for later use. As a bird watcher, Nicola Clayton had always thought that these birds must have some kind of memory of past events, what we call episodic memory. She had observed these birds hiding perishable foods—lots of it—and remembering where to retrieve it and when. At the time, in the 1990s, it was thought that this kind of memory was the prerogative of primates. At a conference, Clayton got talking to the Cambridge animal behavior expert

Anthony Dickinson. He first dismissed the idea of episodic memory in birds. But when Clayton told him about the caching behavior, he was intrigued. They decided to collaborate.

In 1998 Clayton and Dickinson published a paper in the prestigious journal *Nature*, detailing the memory capacities of jays in their stashing behavior. A pillar of primate uniqueness had fallen. Clayton now runs the comparative cognition laboratory at the University of Cambridge and has demonstrated many of the behaviors associated with Jay's caching behavior in controlled laboratory experiments. These include not only its episodic memory, but also behaviors such as predicting what it needs in the future and understanding the perspective of other thieving jays. I remember Clayton talking about these studies when I was a postdoc in London, it was one of my first encounters with comparative research that got me interested in brain evolution.

Sophisticated bird behavior has now been observed in many instances, both in the lab and in the wild. One of the most famous examples is the opening of milk bottles by tits in the United Kingdom in the 1920s. Milk bottles left in front of doors had a foil cap. One clever tit discovered that if it pierced the foil with its beak it could just reach in and drink some of the fatty top layer of the milk. Soon other birds started copying this behavior. Before long, it had spread across the country and it went on for 20 years.

These birds, apparently, were learning from each other. This is another behavior we would associate with clever primates, or at least with mammals, with their extensive six-layered neocortex, but not with tiny bird brains. How much learning precisely happens in this situation has led to fierce debate among animal psychologists, but it has now been shown that these birds can learn new behaviors and transmit them throughout their social network. In primates, we would call such behavior "cultural transmission."

But how are these sophisticated behaviors possible with a brain weighing less than 10 grams? Mammals have invested millions of years of evolution in a large, six-layered neocortex that consists of sophisticated processing units. The bird brain was always assumed to be too small, too primitive, too simplistic. We now know that all three of these assumptions were wrong. But to better understand the bird brain, we need to first understand bird evolution.

One of the proposals of Darwin's theory of evolution by natural selection is that there should exist fossils of hybrid species—species that are not quite like the common ancestors of two present-day species, but also not quite the full products we know now. The "missing link" between apes and humans that Eugene Dubois searched for in chapter 1 is an example. So is the *Tiktaalik* Neil Shubin discovered, that showed the transition from sea to land described in chapter 2.

Another person always on the lookout for possible missing links was Thomas Henry Huxley. In the late nineteenth century, he was one of the fiercest defenders of Darwin's new theory of evolution and natural selection, styling himself as "Darwin's bulldog." When Huxley got wind of a fossil found in Bavaria, he knew he had found a missing link. The fossil had many hallmarks of a reptile—such as a long tail and sharp claws—but it also had a wishbone and beautifully preserved wings and feathers. It looked like a reptile-bird hybrid. It also bore similarities with other skeletons found around the same time, skeletons that clearly belonged to the greatest of all reptiles, the dinosaurs. Based on the hybrid, Huxley proposed that birds evolved from dinosaurs. Evolutionarily speaking, he proposed that birds *are* dinosaurs.

The consequences of Huxley's proposal are far reaching. It means dinosaurs did not all die when the meteorite hit 66 million years ago. Some small specimens survived. They became birds. What is left of the mighty *T. rex* that hunted our poor early mammalian ancestors is the chicken on your plate. For a long time, Huxley's theory remained dormant. Feathers don't fossilize well, and most dinosaur skeletons were perfectly interesting without speculations about feathers. But starting in the late 1990s, fossils from the Liaoning province of northeastern China came to light. The circumstances in China at the time of the dinosaurs featured a forest-rich landscape occasionally plagued by volcanic eruptions. Those eruptions tended to bury and preserve any animal they overtook, including any feathers they might have.

In the fossil beds of Liaoning province, feathers are abundant. They are not a feature of an evolutionary weirdo dinosaur, but a prominent feature across many. Feathers probably did not start out as instruments of flight, but of warmth. Early feathers did not form sophisticated layers on wings as birds have now, but appeared as small hairlike tufts. Wings might not have evolved originally for flight either, but rather as items of display. There can no longer be any doubt, if we want to understand bird evolution, we have to follow the trail from early amniotes, via reptiles, to dinosaurs and birds.

As we saw in chapter 2, after the great Permian extinction event 252 million years ago, the amniote lineage that led to reptiles became dominant. Reptiles split into two lineages. The lepidosaurs ("scaly lizards") went on to form lizards and snakes. The archosaurs ("ruling lizards") went on to become crocodiles, dinosaurs, and birds. During the 150 million year period leading up to the unfortunate collision with the meteorite, dinosaurs ruled the earth, whether it was terrestrial, aquatic, or aerial niches.

The flying reptile pterodactyl is known to most children from pictures in their story books, but it is not an ancestor of birds. Birds, surprisingly, derived from bipedal meat-eating dinosaurs. They are part of the dinosaur family of therapods, together with *T. rex* and *Velociraptor*. Therapod fossils show many adaptations we now associate exclusively with birds. In fact, many aspects of bird anatomy seem present in some groups of therapods, often for reasons completely unrelated to flight. Hollow bones were always thought to be a crucial adaptation to save weight during flight, but they appeared 100 million years before birds as part of an efficient respiratory system allowing oxygen exchange with the blood during both in- and exhalation. This system likely facilitated the energetic lifestyle of a terrestrial predator. The wishbone, a fusing of the collar bones, is a quintessential bird adaptation that helps produce force during flight, but its original function was probably related to absorbing shocks related to grabbing prey. Some of these therapods became smaller, allowing them to exploit a new niche, that of the trees. From there, some probably started to experiment with hopping between tree branches or gliding between trees. Their feathers and elongated arms might have helped with that. Before the meteorite hit, the first real birds capable of powered flight probably appeared—not as a result of careful engineering design, but as a hodgepodge of adaptations that later proved useful for flight.

What happened to the brains of these animals? As we saw in chapter 2, amniotes followed the pattern of vertebrate brains, but with quite a few changes related to their terrestrial niche, which brought challenges of moving and sensing in quite a different way from their ancestors. Brain expansion probably occurred at multiple stages in the lineage leading up to birds. Predators often have bigger brains than their prey, so it is likely the family from which birds evolved were some of the brainier predatory dinosaurs. The forebrain, especially, seems to have increased in predator dinosaurs. In the bird lineage, bodies also tended to get smaller, probably to help the arboreal lifestyle. Relatively speaking, the brain got bigger. So not only did the brain increase in size, but it increased significantly relative to the rest of the body.

The proto-bird brain was bigger, but still quite small. It has to be, if we expect a bird to fly with it. That has always been the assumption at least—that the size of the bird brain is fundamentally constrained by the need to carry it around in the air and that this puts a fundamental limitation on bird brain power. But new experiments are showing that birds may be much smarter than we could have predicted. To see how clever the bird brain is, let's turn to Suzana Herculano-Houzel's brain soup one more time. With her collaborators, she investigated how many neurons the tiny bird brain can pack.

Remember that primates are exceptional in packing more neurons in the same amount of brain tissue. It turns out birds are even more extreme. Herculano-Houzel and her colleagues compared brains from mammals and birds of the same weights and consistently found that the birds packed many more neurons into the same amount of tissue. Take the humble starling. Its brain weighs about 1.86 grams, but it packs 483 million neurons. The brain of the marmoset monkey is more than four times as big, but only packs 1.3 times as many neurons.

Like those of primates, birds' neurons are packed most densely in the telencephalon, the part of the brain containing most of the subdivision we have discussed in the previous chapters. Although they are all pretty impressive, not all birds are the same. Emus and chickens are not the brightest stars in the bird universe, but songbirds and parrots in particular stand out.

There seems to be some convergence between the brains of birds and primates. Both pack a lot of neurons into a much tighter space than most other animals. This gives the brains of both lineages a lot of computing power.

Still, bird brains and mammalian brains look very different. At the beginning of the twentieth century, the comparative anatomist Edinger proposed that the forebrain of the bird brain consisted mostly of areas similar to the subcortical part of the mammalian brain. The mammalian neocortex with its regular six layers has no equivalent in the bird, he proposed. This was the dominant view for the next century, even as more and more evidence started accumulating that was incompatible with this view. Eventually, the neuroscientist Eric Jarvis and his colleagues convened a series of meetings inviting a group of world leading experts on the bird brain to reassess similarities and differences between the mammalian and the bird brains.

It was not an easy task. Put two scientists together in a room and they're bound to disagree on everything—put 30 together and ask them to revise 100 years of orthodoxy and there will be chaos. One of the attendees referred to the endeavor as "herding cats." But eventually a consensus emerged. The revised picture of the bird brain was very different from Edinger's proposal. Instead of the idea that most of the bird brain is devoted to the equivalent of the mammalian subcortex, the experts proposed that the majority of the forebrain is homologous to the mammalian neocortex. It is not a six-layered neocortex, as in the mammal, but it has the same origins, in the amniote dorsal pallium.

This has been the new consensus view ever since. It appears that birds have invested in an extended dorsal pallium, just as mammals have. Birds also found a way to pack more neurons into a small space, just as primates have. But one mystery remained.

The proposed advantage of the neocortex had always been thought to be its very regular, column-wise organization. This produced a great architecture for efficient computation, with information exchange across the layers of the cortex to create the processing unit of the cortical column, and between the columns to integrate across the brain. The bird brain, in contrast, is organized in a nuclear fashion, with groups of neurons clustering together. This does not lend itself to the building of repetitive processing units, but it turns out that some principles of the cortex are mimicked in the bird brain.

The bird brain is a bit too small for MRI, as we did in the primates. Therefore, in 2020, Onar Güntürkün and his colleagues decided to use a more sophisticated technique, called polarized light imaging. This technique relies on the refraction of light in very thin slices of brain tissue. This refraction is influenced by the presence of white matter fibers. This way, they could visualize how the bird brain is wired, just as we did with MRI for the primate brain. They saw that a structure similar to the neocortex, with connections forming columns and connections across areas to allow integration of information, was present in some parts of the bird brain. Thus, although the bird brain might not have the big, flexible association cortex of the primate, it not only mimics dense packing of neurons and expansion of the forebrain, it also mimics some of the ways to organize the forebrain. These analogous structure show that, to some extent, the primate mammal and the bird converged on similar solutions despite very different starting points.

The bird has always been a prime example of convergent evolution because of flight. Flight evolved only three times in vertebrates: in bats, pterosaurs, and birds. Each of these found a way to convert their front limbs in such a way as to enable flight. Their solutions aren't exactly the same. In bats the bones of the five digits are enlarged to allow flexibility and independent movement of the wings, whereas in birds there is no independent digit movement anymore. The importance, though, is that they started out with a different body and ended up with a similar solution. Convergent evolution can occur anywhere in a species' natural history, from movement patterns to the brain.

There remains a lot to be discovered about the bird brain. Indeed, it never ceases to amaze me that as soon as we move beyond the human brain and its favored "models" such as the macaque monkey, we know so little about the brains of other species. But what we now do know about the bird brain suggests an impressive example of convergent evolution. Starting out with a reptile brain, birds invented a way to pack many neurons into a small space, to create an efficient complex processing unit, and to exchange information across units just as mammals did. The bird brain is capable of a series of complex behaviors. While we might never understand the inner mental life of birds, their behavior is at least suggestive of memory, mental time travel, and some theory of mind reminiscent of that of our own primate brain.

———

Places and grids

Most of the foraging described in this book relies on navigation. Whereas the adult sea squirt was content to find a spot to settle and wait for nutrients to appear, early vertebrates with their predatory lifestyle needed to be able to find their way outside the direct and familiar surroundings. They developed a system capable of navigating new territory. This is where the hippocampus, an ancient part of a three-layered cortex, takes center stage.

In the middle of the twentieth century, the most common theory was that animals navigate by learning combinations of stimuli and responses. Behaviorists performed experiments in which they put animals, mostly rats, in a simple maze, often in the shape of a T. If they put the rat in the bottom arm of the T and then put food on one side of the T but not the other, the rat would quickly learn to move to the correct side. The common interpretation was that the rat learned a simple stimulus-response relationship to solve the task. Upon encountering a particular stimulus—the T junction—the rat learned that a particular response—turn right—would lead to a reward. Indeed, behaviorists went on to show that many complicated behaviors of animals can be reduced to learning lots of stimulus-response relationships. Animals, they argue, learned the simplest rules to achieve their goals.

Some, however, did not agree. The psychologist Edward Tolman was one of them. In his experiments, rats were placed in a maze with lots of environmental cues around the maze that they could use for their navigation, such as lights, objects, and pictures on the wall. The rats were trained to learn to associate one arm of the maze with a food reward, just as before. Then, after training, the rat was placed in another part of the maze to start. If the rat had simply learned to associate the environmental stimuli with a specific response ("Go left if you see the junction with the blue light"), they would not find the reward. But what Tolman found was that the rats were capable of finding their way. Subsequent studies showed that rats could even find shortcuts in mazes or find new routes when old ones were blocked. These results can only be explained, Tolman argued, if the rat possesses a map-like representation of the maze instead of just learning stimulus-response combinations.

This "cognitive map theory" received a major boost in the 1970s, when John O'Keefe and his colleagues started recording the electrical activity of neurons in the hippocampus while rats were freely moving around. The hippocampus is part of the medial pallium that also has a homologue in reptiles. This means it is likely that something similar existed in the amniote ancestor of us all. If you cut through the hippocampus at a certain angle, the section looks a bit like a seahorse, *hippocampus* in Latin, hence the name.

Recording from this area, O'Keefe noticed that certain cells would fire whenever the animal was in a certain part of the room. Other cells would fire when the animal was in a different part, and so on. Different cells, in other words, had different firing locations. They termed these cells "place cells." O'Keefe and his colleague Lynn Nadel argued that through these place cells the hippocampus houses an internal map of the world, just as Tolman had predicted. In subsequent years, a whole plethora of cells have been discovered in the hippocampus, including cells that code head direction, the edge of explored space, and timing. All these cells code information highly relevant to navigation.

The role of the hippocampus in navigation is now quite clear. It is the basis of the result we discussed in chapter 5, where we saw that London taxi drivers have a bigger hippocampus than non-taxi drivers. These effects are causal. London taxi drivers train for years, memorizing the more than 26,000 streets in the city and practicing to quickly determine the fastest way between any two random points. When researchers tracked the size of the hippocampus in apprentice taxi drivers over the years of training, they saw that their hippocampus increased in size over time. Moreover, students that failed and never made the cut didn't show the same increase.

Place cells   Grid cells

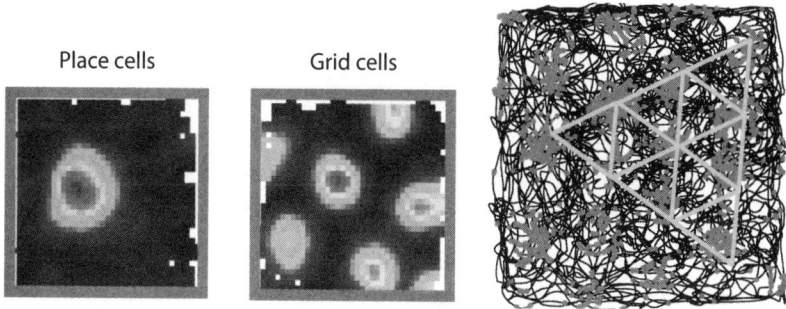

FIGURE 7.1. Place cells and grid cells. Figures on the left and in the
middle show a map indicating where a particular neuron fires. A place
cell fires when a rat moves through a particular part of space. A grid cell
fires in multiple places. On the right, we see the trajectory a rat took on
the map and where the particular cell fired. The equilateral triangles
illustrate that these locations are organized in a hexagonal pattern.

This shows us that the size of the hippocampus is not an
inborn characteristic—it's not that a bigger hippocampus pre-
disposes you to become a taxi driver—but it is related to your
progress, or lack of it. Another interesting effect that relates to
the patterns observed in the hippocampal place cells is the co-
driver effect. Have you ever noticed that you do not learn a route
that well if you are the passenger compared to when you do the
driving yourself? The hippocampal place cells of rats learn a lot
faster when the rat is moving around itself as well. Our navigational system has many similarities to that of the rat, due to our
homologous hippocampus.

Interesting as these results are, the main reason researchers
were interested in the hippocampus had, at first sight, little to
do with navigation. They were interested in memory. The role
of the hippocampus in memory cannot be illustrated more dramatically than in the case of a patient known for years by the
initials H.M. From early adolescence, H.M. suffered from

severe and uncontrollable epilepsy. After all other treatments failed, doctors tried to alleviate the epilepsy by removing the hippocampus and surrounding areas in both hemispheres of H.M.'s brain. The results were dramatic.

Following the surgery, H.M. was no longer able to store any new explicit memories. If you would tell him a joke, he would forget it quickly. If he had met somebody new, he would greet them again as a total stranger the next day. He was able to hold information for 30 seconds or so, but committing it to long-term memory was impossible. He was able to remember things from before surgery, and he was able to learn new skills that do not rely on explicit recall, but consolidation of conscious memories was totally abolished. As described by one of his biographers, he lived in a permanent present tense. Somehow, this system that evolved for navigation in early vertebrates seems to have a major role in the ability that is most closely linked to sense of self: our ability to consciously remember our past.

How the two functions of the hippocampus, spatial navigation and memory, are related remained a mystery for a long time. The two strands of research moved on more or less in parallel, although it was known that space and memory are linked in some way. When we describe a memory we often use spatial terms. In Alzheimer's Disease, a debilitating dementia in which patients have severe explicit memory problems, getting lost is a prominent early symptom.

As researchers continued to probe the place cells of the hippocampus, they were starting to see things that suggested a role in memory. Place cells, for instance, do not only code the place an animal is currently occupying. Some represent places the animal has just visited, representing the recent past instead of the present. If a rat was lost, the cells representing the place where it thought it was fired, a result dependent on a memory

of previous experience. Cells also fired in response to nonspatial cues, such as odors or visual information, and then fired at the
location where that stimulus was encountered, suggesting they
have a role in associating locations with events that took place
there.

Intriguingly, waves of activity were detected in the hippocampus of animals that were asleep or resting. Interfering with
this activity resulted in the animals not consolidating whatever
route or associations they had been taught earlier. This activity
during sleep seems important in consolidating the information
from the hippocampus to more permanent storage in the neocortex. Waves of activity in the hippocampus are not random.
They are seen when animals plan a route in a known map, starting with cells representing the current location and ending in
the goal location. Space, memory, and mental time travel thus
seem to be part of a unified set of processes centered around the
hippocampus.

One question that kept mystifying researchers is where the
place cells in the hippocampus get their information. In the beginning of the century, Edward and May-Britt Moser started to
explore this question. They first demonstrated that a part of the
hippocampus seems to contain place cells even if it does not
receive any information from the rest of the hippocampus. The
information had to come from another structure. Together with
an anatomist, they carefully mapped out where this part of the
hippocampus got its inputs. This turned out to be a nearby
structure, the entorhinal cortex. They started recording from
this part of the brain to see if any cells there responded to spatial
locations.

They found cells in the entorhinal that fired at particular locations, but in a different way than the hippocampal place cells.
These entorhinal cells fired at multiple locations in space, with

clear areas of silence in between. When they expanded the spatial environment in which the rat could move and visualized the places where these cells fired, they saw they formed a neat hexagonal pattern, a spatial grid. Now, any mathematician will tell you that if you want to represent a 2-D space in an efficient manner, a hexagonal grid pattern is the way to go. The entorhinal cells encoded space in such an efficient pattern.

These "grid cells" were very stable. Whereas hippocampal place cells easily adapt to a new environment by changing their firing pattern, the grid cells—as well as other cells detected in the entorhinal cortex representing borders and head directions—were remarkably stable. One part of the system provides an overall way of describing the spatial organization, the other part a flexible map dependent on the context. Together, the entorhinal-hippocampal system forms a system with multiple representations of space that can order information and consolidate it into long-term memory. O'Keefe and Edward and May-Britt Moser were awarded the Nobel Prize in Physiology or Medicine for their discoveries of place and grid cells.

If this system is so efficient in representing and ordering spatial information, and we believe this forms the basis of our explicit memory system, can this system represent nonspatial information as well? In our human brain, the entorhinal-hippocampal system is connected to the large and expanded neocortex, which theoretically could provide it with all sorts of information. Could the same system organizing and storing spatial information be used for much more abstract information?

This was the question asked by Tim Behrens, Jill O'Reilly, and their student Alexandra Constantinescu in a paper in 2019. At the time, it was already known that the spatial hexagonal grid could be detected in humans using functional neuroimaging. Although it is impossible to record from single cells in the entorhinal cortex in humans, you can measure whether cells respond in a grid-like manner by asking people to move in a direction that is aligned to the grid or not. If the movement is aligned with the grid, cells fire more often and you will detect more signal than when they are moving in a direction not aligned with the grids.

In the 2019 study, instead of moving through a spatial environment, the participants had to navigate an abstract space. Pictures of birds were constructed with different lengths of the neck and legs. Some birds were associated with a particular outcome. Participants were exposed to particular neck:legs ratios, in essence learning to navigate a "bird space." During the experiment, participants saw birds morphing for a while and were then asked to imagine the bird that would appear if the morphing continued. In other words, they had to imagine traveling through bird space. The experiment was designed so that some of the imagined morphing was along the grid of the learned bird space and some was not. Lo and behold, traveling along the grid of bird space was associated with greater signals in the entorhinal cortex, suggesting we humans can use this old system for spatial navigation to also help us move through abstract, conceptual spaces.

Subsequent studies have shown that the human entorhinal-hippocampal system is involved in ordering many different types of information. Humans use cognitive maps in abundance, whether for spatial navigation, bird space, odors, or sounds. It also works for social information. If people are shown pairs of relationships across a group of people, they will spontaneously organize the information in a "social map" and travel along the map if they want to find out the relationship between any two people they have not previously seen together. The grid-like code is not limited to the human entorhinal cortex, but can also be observed on regions of the neocortex traditionally associated with autobiographical memory.

The extended human neocortex is in constant interaction with the evolutionary ancient hippocampus-entorhinal system, providing it with new types of information and using its outputs for further analysis. As such, the system gets retrofitted for additional functions.

————

A reading brain

The past chapters have charted how different branches of the animal tree of life adapted their brains to suit their needs. Through slow modification by slow modification, different species ended up with different brains that produced a behavioral repertoire that allowed them to deal with the challenges they faced when finding food and staying alive. It would be tempting to conclude that each aspect of behavior has some corresponding detail associated with it in its owner's brain. But our human brains are capable of functions that evolved too recently to be the result of major changes in the brain itself. How did these functions come about?

Take reading of this sentence. Reading (and of course writing) was invented in Mesopotamia about 5500 years ago and again independently in Egypt, China, and the Americas. Early writing consisted of pictograms that represented whole words or concepts. Egyptian hieroglyphs are an example. The idea of an alphabet is to use a limited set of characters that represent sounds and combine to form words. This idea was invented only once, about 2000 BC. The alphabet used to write this sentence is commonly referred to as the Latin alphabet and appeared around 700 BC, but it has undergone many changes since. The distinction between lower and upper case letters, for instance, is a Medieval invention.

Literacy itself was confined to a small percentage of the population until fairly recently. According to the World Bank, in 2020 about 90% of people over 15 years of age worldwide could read. Two hundred years earlier, in 1820, literacy was estimated to be only 12%. It is unlikely we have a brain system that evolved specifically for a behavior that until recently most of us did not display. How it is possible for this kind of behavior to emerge is a question French cognitive scientist Stanislas Dehaene has spent a large part of his career investigating. His answer is yet another example of retrofitting the brain—this time by culture.

In the previous chapter, we saw that humans are very good at passing on information to the next generation to build on. Knowledge accumulates across generations, and this is what happens in the emergence and continuous transformation of our reading and writing abilities. We call this culture. One way to explain this ability is to postulate that our human brain processes a general "cultural module" that is important for this function. We have already seen that large parts of the human neocortex are dedicated to processing and reprocessing of increasingly abstract information, so perhaps there is a general module at the top of all this that can bring everything together and allow us our cultural abilities. Such a module would make localizing any cultural behavior in the brain quite easy; it should always involve the cultural module.

To test the notion of a cultural module, we can compare brain activations related to reading with that of another cultural skill, mental arithmetic. Like reading, arithmetic is a skill we all have to learn during our lifetime and is the result of many generations of accumulated knowledge. Our system of using the position of a limited set of symbols to indicate different values was probably invented independently in Mesopotamia, China, and the Americas.

If we ask people to perform mental arithmetic, one prominent region that shows an increase in activation is in the parietal cortex, at the top of the brain. Recall we discussed this part of the brain in chapter 3, as part of the stream controlling visually guided reaching and grasping movements. In contrast, during reading, areas lower down in the brain show increased activation. These are areas that are part of the temporal visual stream from chapter 4. There does not seem to be a single repository for cultural behavior in humans. Rather, different cultural behaviors rely on different systems. Interestingly, while reading and arithmetic are located in very different parts of the neocortex, they both show very little variability in their location across individuals. Different places for different functions, but always the same place for any given function.

Let's look at some of the regions involved in culturally transmitted behaviors a bit more closely. Reading activates a group of regions in the left hemisphere of the brain. A number of these regions are important for language in general, a skill that did evolve over a long enough time to have dedicated regions evolve, but one temporal visual stream region seems particularly active in response to the written word. It responds to letters, but not to equally simple shapes or more complex shapes such as houses or faces. Dehaene and his colleagues called it the visual word form area. The activation of the region is related to your expertise in reading. It is more active in response to familiar than less familiar script. It is also more active to familiar letter combinations than to unfamiliar combinations. When human volunteers were asked to read letter strings that differed in how well they obeyed the rules of their language, the more the string followed the rules, the more activation the region showed.

To really investigate the function of the visual word form area in reading, it would be ideal to compare the activity of the region between literate and illiterate people. Illiterate communities are becoming harder and harder to find. When they do exist, they are unlikely to live near an MRI scanner. Dehaene and colleagues managed to study 63 people with a range of literacy, but otherwise from the same communities in Brazil and Portugal. Some had not attended school as a child and were illiterate, some had not attended school but had learned to read later in life, and some had attended school and were literate. The visual word form area was more active in response to words and sentences in the literates, whether they learned early in life or later. Moreover, how active the region was could be predicted by an individual's reading performance. All this suggests a region that is involved in reading as a learned skill.

In a subsequent study, Dehaene teamed up with another French researcher, Michael Thiebaut de Schotten, to study how the visual word form area is connected to the rest of the brain. They saw that not only did the visual word form area become more active as a function of literacy, it also changed its connections. Its connection to the major white matter pathways of the human language system become much stronger after learning to read. The region got plugged into the language system, but only in people who learned to read.

There must be something in this region that makes it suitable to be co-opted by reading when the cultural circumstances are right. It must have an original, let's call it hardwired, function that makes it useful for later reading. It does seem that this region codes a particular combination of lines and angles. That would make sense, to identify words, it helps to be able to detect combinations of lines and angles that form letters or other characters. But the reason this region so easily gets involved in reading is only partly because this region responds to the right things; "the right things" are also shaped that way because of the brain.

As we discussed before, the primate brain is a visual brain. Our neocortex contains dedicated visual pathways to deal in particular ways with visual information. The parietal stream is important for controlling the motor system, and the temporal stream is important for setting the sensory context. This last function includes processing objects, be it faces for our elaborate social life or objects to judge their suitability in a foraging situation. Although processing a complex visual scene full of overlapping objects seems natural to us, we rely on quite a few primate specializations to accomplish it.

For instance, one important function our visual system has to perform is to segment the visual scene into distinct objects. This can be done by identifying discontinuities in the visual science. A change in color or texture might indicate the border of an object. However, this can be an unreliable cue, as some objects do not differ from their background. This is exploited by animals using camouflage. Primates therefore rely on an additional cue. Our two frontally oriented eyes pick up a slightly different version of the 3-D scene, and that means they see more or less of a partially occluded object. We can use this information to infer which part of the scene belongs to which object. In a recent experiment, it was demonstrated that mice do not use this extra source of information. Our temporal visual stream really processes the world in a very different way from that of other mammals.

Reading and writing rely on our primate specialties. Dehaene refers to the work of the cognitive scientist Mark Changizi, who has argued that all the scripts in the world, no matter how different they appear at first glance, share certain features in how they are constructed. Certain subshapes occur with remarkably similar frequencies across different scripts. What's more, these frequencies in scripts match the frequencies with which they are encountered in natural visual scenes. The way characters are constructed also follows a general pattern. Combinations of three strokes occur very often. Higher up the writing hierarchy, combinations of about three subunits, such as letters forming a morpheme, also occur frequently. Dehaene and colleagues argue that these features are likely due to the brain shaping writing, rather than the other way around. Our visual system evolved for us to look at images of objects, so when we started writing it would make sense we used shapes and forms that we could most easily decipher.

Similarly, remember that the temporal visual stream is hierarchically organized. The more anterior in the temporal cortex we move, the more individual features get combined. This combining often involves an increase of the receptive field of the neuron. This is remarkably similar to the hierarchical organization of our written scripts. Thus, our writing is shaped to be easily processed by the visual system in our brain. In turn, with frequent exposure, our brain is plastic enough to adapt, but rigid enough that this adaptation always takes the same form. The visual regions that are good at recognizing the letters and words then get wired to the language system. Dehaene terms this "cultural recycling" of a brain region. I would call it retrofitting.

Recap: Brain evolution

As we have seen throughout this book, there is a diverse evolutionary tree that illustrates the path from the common ancestor of all vertebrates to all the vertebrates alive today. At every junction, different circumstances drove different individuals to choose a different path. Multiple species formed, each with their own distinct body and brain.

Of course, "circumstances" in this context suggests maybe a more ordered process than it really is. There is a lot of chance in how the environment shapes up, even more chance in what genetic variations are around in a population at the time, and finally there is the more or less orderly process by which individuals with certain characteristics that result from those genetic variations become more frequent in the population. There are laws at work, first formulated by Charles Darwin and later amended and integrated with an emerging understanding of population genetics and, more recently, with a new understanding of development. These are laws of probability, providing a likelihood of outcomes, rather than discrete predictions. Working on many chance events, these laws led to the brains we see in nature today. This means that although we cannot predict the course of evolution precisely, the diversity we see in nature is constrained enough for the comparative method to work—otherwise this book would not have been possible.

In the last sentence of *The Origin of Species*, Darwin describes the process of evolution as leading to "endless forms most beautiful." Vertebrate brain evolution indeed led to a fantastic variety of brains. Outside the vertebrates, there are even stranger brains of which we have almost no understanding. The octopus brain and the brain of the bee are just two examples that have received some attention in the popular press recently. But the laws of evolution mean that variety cannot be endless. Convergent evolution such as between birds and primates, neural upgrading by teaming up the ancient vertebrate hippocampus with the expanded neocortex, and neural recycling of areas in the visual streams are three ways in which evolution retrofits the brain.

Coopting the parietal visual stream for domain-general intelligence (chapter 3) and the temporal visual stream for processing social information (chapter 5) are others. They illustrate how we can understand a brain only in light of constraints. We need to understand how a working brain is modified into another working brain at each step of evolution. In that process, the new brain by necessity retains almost all the characteristics of its ancestor. It is why our fornix is curled around our thalamus, why most of our conscious processing is visual, why we talk about memory in spatial terms, and why the tools we use and the abstract concepts we generate are all remarkably consistent across human populations. The marks of our evolutionary past determine how we experience our lives.

Darwin might have overstated the case when he observed that evolution leads to endless forms most beautiful. They are indeed beautiful. They are also extremely diverse. But not endless. That is what make brain evolution fascinating. And it explains how what is essentially an organ for helping an animal forage successfully is involved in so many complicated and wonderful behaviors.

ACKNOWLEDGMENTS

Science is teamwork for introverts. This book builds on the work of many scientists, some personally known to me, but many, many whose work I know only via the literature. In my own scientific life, I have always been lucky enough to have fantastic mentors, including my PhD supervisors Ivan Toni, Mike Coles, and Wouter Hulstijn. Ivan in particular is to blame for my interest in biology. During my postdoctoral career my mentor was Matthew Rushworth, who gave me the chance to look at other brains and taught me the value of the long view of understanding the brain. Independent researchers don't exist, so in my "independent" career I've still had the benefit of mentors, including Heidi Johansen-Berg and Pieter Medendorp.

My ideas about what I tend to call neuroecology have been stimulated, challenged, discussed, and refined through my interactions with Dick Passingham. His work and that of his friends Steve Wise, Betsy Murray, and others has been tremendously influential in my work. I hope they feel I have done their work justice, even if they do not agree with all my conclusions. Robin Dunbar has also always been extremely supportive, from the moment I sent him a naive email: "we're doing social stuff too."

The members, past and present, of the Cognitive Neuroecology Lab never cease to amaze me in their enthusiastic tolerance of my crazy ideas and, more importantly, never cease to surprise me with their own. Many of them also helpfully went through various sections of the manuscript and provided comments, as did Jim Rilling, Leah Krubitzer, and Matthew Rushworth. My many colleagues and scientific friends, including prominently Saad Jbabdi, Jerome Sallet, Sasha Khrapitchev, and MaryAnn Noonan, make my work both better and much more enjoyable.

Alison, Hallie and Laura of Princeton University Press provided both excellent editing and council, guiding a first-time book writer through the process.

I have been blessed with some very good friends on the journey through life, of which Patrick, Robartus, and Johannes deserve particular credit for keeping me at least somewhat sane. Finally, I wish to thank my wonderful family. My parents, who started my interest in the world around me. And of course Jill and our three children, Kitty, Johannes, and Tessa—probably the only children ever to ask, "Dad, can we look at brains on your phone?"

FURTHER READING

This is an inexhaustive list of works I used in writing this book. I have tried to include a selection of generally accessible books for those who want to dive deeper.

Introduction: Different brains

Evolution by natural selection is of course best described by Darwin in The Origin of Species *(1859), and his view on the evolution of humans is described in* The Descent of Man *(1871). The development of Darwin's thinking is described in the excellent biography by Janet Browne. The life of Eugene Dubois is described by Shipman (2001).*

Browne J (2003) *Charles Darwin. Voyaging.* London: Pimlico

Browne J (2010) *Charles Darwin. The power of place.* London: Pimlico.

Darwin C (1859) *On the origin of species by means of natural selection, or the preservation of favoured races in the struggle for life.* London: John Murray

Darwin C (1871) *The descent of man, and selection in relation to sex.* London: John Murray

Mars RB, Foxley S, Verhagen L, Jbabdi S, Sallet J, Noonan MP, Neubert FX, Andersson JL, Croxson PL, Dunbar RIM, Khrapichev AA, Sibson NR, Miller KL, Rushworth MFD (2016) The extreme capsule fiber complex in humans and macaque monkeys: A comparative diffusion MRI tractography study. *Brain Structure and Function* 221:4059–4071

Shipman P (2001) *The man who found the missing link: The extraordinary life of Eugene Dubois.* New York: Simon & Schuster

Chapter 1. The sea squirt, the lamprey, and the bee

The example of the sea squirt and movement as the brain's main role was first described in Llinás's I of the Vortex. Cooking as a way to outsource food processing to support a bigger brain is an idea advanced by Wrangham (2009). Evolution of brains up to early

vertebrates is covered in depth by Striedter and Northcutt (2020). The idea of the brain as a foraging system—with a strong emphasis on primate prefrontal cortex—is developed at book length by Passingham and Wise (2012).

Anderson PR, Woltz CR, Tosca NJ, Porter SM, Briggs DEG (2023) Fossilisation processes and our reading of animal antiquity. *Trends in Ecology and Evolution* 38:1060–1071

Cooper GM (2000) *The cell: A molecular approach (2nd edition)*. Sunderland: Sinauer Associates

Costa KM, Schoenbaum G (2022) Dopamine. *Current Biology* 32:R807-R827

Glimcher PW (2011) Understanding dopamine and reinforcement learning: The dopamine reward prediction error hypothesis. *Proceedings of the National Academy of Sciences USA* 108:15647–15654.

Hebb DO (1948) *The organization of behavior. A neuropsychological theory.* New York: Wiley and Sons

Kristan WB Jr (2016) Early evolution of neurons. *Current Biology* 26:R937-R980

Llinás RR (2001) *I of the vortex: From neurons to self.* Cambridge: MIT Press

Martin W, Baross J, Kelley D, Russell MJ (2008) Hydrothermal vents and the origin of life. *Nature Reviews Microbiology* 6:805–814

Montague PR, Dayan R, Person C, Sejnowski TJ (1995) Bee foraging in uncertain environments using predictive Hebbian learning. *Nature* 377:725–728

Montague PR, Dayan P, Sejnowski TJ (1996) A framework for mesencephalic dopamine systems based on predictive Hebbian learning. *Journal of Neuroscience* 16:1936–1947

Murray EA, Wise SP, Rhodes SEV (2011) What can different brains do with reward? In: Gottfried JA (Ed.) *Neurobiology of Sensation and Reward.* Boca Raton: CRC Press

Nobel Prize Outreach (1973) *Press release. The Nobel prize in physiology or medicine 1973.* https://www.nobelprize.org/prizes/medicine/1973/press-release

Northcutt RG (2012) Evolution of centralized nervous systems: Two schools of evolutionary thought. *Proceedings of the National Academy of Sciences USA* 109:10626–10633

O'Reilly JX, Mesulam MM, Nobre AC (2008) The cerebellum predicts the timing of perceptual events. *Journal of Neuroscience* 28:2252–2260

Passingham RE, Wise SP (2008) *The neurobiology of the prefrontal cortex: Anatomy, evolution, and the origin of insight.* Oxford: Oxford University Press

Ryan K, Lu Z, Meinertzhagen IA (2016) The CNS connectome of a tadpole larva of *Ciona intestinalis* (L.) highlights sidedness in the brain of a chordate sibling. *eLife* 5:e16962

Schultz W, Dayan P, Montague PR (1997) A neural substrate of prediction and reward. *Science* 275:1593–1599

Striedter GF, Northcutt RG (2020) *Brains through time. A natural history of vertebrates.* Oxford: Oxford University Press

Wrangham R (2009) *Catching fire: How cooking made us human.* London: Profile Books

Chapter 2. The dragon and the shrew

*The triune brain theory in its most extensive form is described by MacLean (1990). It was
popularized by Carl Sagan in his 1977 book* The Dragons of Eden. *The discovery of
sea-to-land Tiktaalik is described by Shubin (2008). The end-Permian and K/T extinctions and the history of our understanding of these events, including a graphic description
of the end-Permian extinction through the eyes of an unlucky herbivore, are described in
Benton (2015). The search of the cause of the dinosaur extinction is described first-hand
by Alvarez (2015). The idea of medial frontal cortex supporting foraging decisions is based
on Murray et al. (2011).*

Alvarez W (2015) T. rex and the crater of doom. Princeton: Princeton University Press

Benton MJ (2015) *When life nearly died. The greatest mass extinction of all time.*
London: Thames and Hudson

Chalmers Mitchell P (1927) Reptiles at the Zoo: Opening of new house today. *The
London Times* 15 June 1927, 17

Cooke DF, Baldwin MKL, Donaldson MS, Helton J, Stolzenberg DS, Krubitzer L
(2016) Lab rats gone wild: How seminatural rearing of laboratory animals shapes
behavioral development and alters somatosensory and motor cortex organization.
Society for Neuroscience Abstracts

Frith CD, Friston K, Liddle PF, Frackowiak RSJ (1991) Willed action and the prefrontal cortex in man: a study with PET. *Proceedings of the Royal Society London
B* 244:241–246

Kaas JH (2011) Reconstructing the areal organization of the neocortex of the first
mammals. *Brain Behavior and Evolution* 78:7–21

Kolling N, Behrens TEJ, Mars RB, Rushworth MFS (2012) Neural mechanisms of
foraging. *Science* 336:95–98

Kolling N, Wittmann M, Rushworth MFS (2014) Multiple neural mechanisms of
decision making and their competition under changing risk pressure. *Neuron*
81:1190–1202

Krubitzer L, Manger P, Pettigrew J, Calford M (1995) Organization of somatosensory cortex in monotremes: In search of the prototypical plan. *Journal of Comparative Neurology* 351:261–306

Krubitzer L (2007) The magnificent compromise: Cortical field evolution in mammals. *Neuron* 56:201–208

MacLean PD (1990) *The triune brain in evolution. Role in paleocerebral function.* New York: Plenum Press

Murray EA, Wise SP, Rhodes SEV (2011) What can different brains do with reward? In: Gottfried JA (Ed.) *Neurobiology of Sensation and Reward.* Boca Raton: CRC Press

Ouwens PA (1912) On a large *Varanus* species from the Island of Komodo. *Bulletin du Jardin Botanique de Buitenzorg* 6:1–3

Sagan C (1977) *The dragons of Eden. Speculations on the evolution of human intelligence.* New York: Ballantine

Shubin N (2008) *Your inner fish.* London: Penguin

Chapter 3. The squirrel and the squirrel monkey

Suzana Herculano-Houzel tells the story of her brain soup in The Human Advantage *(2016). Milner and Goodale describe their theory of vision for perception and vision for action in their book* The Visual Brain in Action *(2006). The computational properties of the parietal stream are investigated in detail in the fantastically named* Computational Neurobiology of Research and Pointing *(Shadmehr and Wise, 2004). The organization of the motor system according to action categories, as well as a history of studies of the motor cortex, is described by Graziano (2009). A nice overview of research into the brain mechanisms behind intelligence is presented by Duncan (2010).*

Andersen RA, Cui H (2009) Intention, action planning, and decision making in parietal-frontal circuits. *Neuron* 63:568–583

Cartmill M (1992) New views on primate origins. *Evolutionary Anthropology* 1:105–111

Duncan J (2010) *How intelligence happens.* New Haven: Yale University Press

Genovesio A, Wise SP, Passingham RE (2014) Prefrontal-parietal function: From foraging to foresight. *Trends in Cognitive Sciences* 18:72–81

Graziano MS (2009) *The intelligent movement machine. An ethological perspective on the primate motor system.* Oxford: Oxford University Press

Herculano-Houzel S, Catania K, Manger PR, Kaas JH (2015) Mammalian brains are made of these: A dataset of the numbers and densities of neuronal and nonneuronal cells in the brain of Glires, Primates, Scandentia, Eulipotyphlans, Afrotherians and Artiodactyls, and their relationship with body mass. *Brain, Behavior and Evolution* 86:145–163

Herculano-Houzel S (2016) *The human advantage. A new understanding of how our brain became remarkable.* Cambridge: MIT Press

Ingle D (1973) Two visual systems in the frog. *Science* 181:1053–1055

Kaas JH, Gharbawie OA, Stepniewska I (2011) The organisation and evolution of dorsal stream multisensory motor pathways in primates. *Frontiers in Neuroanatomy* 5:34

Karadachka K, Assem M, Mitchell DJ, Duncan J, Medendorp WP, Mars RB (2023) Structural connectivity of the multiple demand network in humans and comparison to the macaque brain. *Cerebral Cortex* 33:10959–10971

Medendorp WP, Heed T (2019) State estimation in posterior parietal cortex: Distinct poles of environmental and bodily states. *Progress in Neurobiology* 183:101691

Milner AD, Goodale MA (2006) *The visual brain in action (2nd edition).* Oxford: Oxford University Press

Mishkin M, Ungerleider LG, Macko KA (1983) Object vision and spatial vision: Two cortical pathways. *Trends in Neurosciences* 6:414–417

Rathelot JA, Strick PL (2006) Muscle representation in the macaque motor cortex: An anatomical perspective. *Proceedings of the National Academy of Sciences USA* 103:8257–8262

Shadmehr R, Wise SP (2004) *The computational neurobiology of reaching and pointing: A foundation for motor learning.* Cambridge: MIT Press

Wise SP, Boussaoud D, Johnson PB, Caminiti R (1997) Premotor and parietal cortex: Corticocortical connectivity and combinatorial computations. *Annual Reviews of Neuroscience* 20:25–42

Wise SP (2006) Ventral premotor cortex, corticospinal region C, and the origin of primates. *Cortex* 42:521–524

Chapter 4: The lemur and the macaque

Evolution of the simian brain in the context of foraging was proposed and worked out in depth in Passingham and Wise (2008) and Murray et al. (2017). The latter's theory is described in a more accessible form in The Evolutionary Road to Human Memory *(2020). Climatic changes and their consequences for flora and fauna are detailed by Prothero (2006). Early primates are described by Fleagle (2013).*

Braunsdorf M, Blazquez Freches G, Roumazeilles L, Eichert N, Schurz M, Uithol S, Bryant KL, Mars RB (2021) Does the temporal cortex make us human? A review of structural and functional diversity of the primate temporal lobe. *Neuroscience and Biobehavioral Reviews* 131:400–410

Bryant KL, Glasser MF, Li L, Bae JJ, Jacquez NJ, Alarcon L, Fields A, Preuss TM (2019) Organization of extrastriate and temporal cortex in chimpanzees compared to humans and macaques. *Cortex* 118:223–243

Fleagle JG (2013) *Primate adaptation and evolution*. Amsterdam: Academic Press

Freedman DJ, Riesenhuber M, Poggio T, Miller EK (2003) A comparison of primate prefrontal and inferior temporal cortexes during visual categorization. *Journal of Neuroscience* 23:5235–5246

Harlow HF (1949) The formation of learning sets. *Psychological Review* 56:51–65

Janmaat KRL, Byrne RW, Zuberbühler K (2006) Primates take weather into account when searching for fruits. *Current Biology* 16:1232–1237

Lambon Ralph MA, Sage K, Jones RW, Mayberry EJ (2010) Coherent concepts are computed in the anterior temporal lobes. *Proceedings of the National Academy of Sciences USA* 107:2717–2722

Luria AR (1966) *Higher cortical functions in man*. New York: Springer.

McKee JL, Riesenhuber M, Miller EK, Freedman DJ (2014) Task dependence of visual and category representations in prefrontal and inferior temporal cortexes. *Journal of Neuroscience* 34:16065-16075

Milton K (1993) Diet and primate evolution. *Scientific American* 269:86–93

Murray EA, Wise SP, Graham KS (2017) *The evolution of memory systems: Ancestors, anatomy, and adaptations*. Oxford: Oxford University Press

Murray EA, Wise SP, Baldwin MKL, Graham KS (2020) *The evolutionary road to human memory*. Oxford: Oxford University Press.

Passingham RE (1982) *The human primate*. New York: W.H. Freeman

Passingham RE, Wise SP (2008) *The neurobiology of the prefrontal cortex: Anatomy, evolution, and the origin of insight*. Oxford: Oxford University Press

Poli F, Ghilardi T, Mars RB, Hinne M, Hunnius S (2023) Eight-month-old infants meta-learn by downweighting irrelevant evidence. *Open Mind* 7:141–155

Prothero DR (2006) *After the dinosaurs: The age of mammals*. Bloomington: Indiana University Press

Seiffert ER (2012) Early primate evolution in Afro-Arabia. *Evolutionary Anthropology* 21:239–253

Van der Linden M, Murre JMJ, Van Turennout M (2008) Birds of a feather flock together: Experience-driven formation of visual object categories in human ventral temporal cortex. *PLoS ONE* 3:e3995

William BA, Kay RF, Kirk EC (2010) New perspectives on anthropoid origins. *Proceedings of the National Academy of Sciences USA* 107:4797–4804

Chapter 5: The fox and the dog

The Siberian fox experiment is described at book length by Dugatkin and Trut (2017). Characteristics of successfully domesticated animals and their effect on human society are described by Diamond (1997). The framework for different types of information from the

eyes of different species is based on work by Perret, Emery, and Mars. The idea of hijacking of the temporal lobe by social information processing is presented by Braunsdorf and colleagues (2021).

Braunsdorf M, Blazquez Freches G, Roumazeilles L, Eichert N, Schurz M, Uithol S, Bryant KL, Mars RB (2021) Does the temporal cortex make us human? A review of structural and functional diversity of the primate temporal lobe. *Neuroscience and Biobehavioral Reviews* 131:400–410

Burger J, Gochfield M, Murray BG (1991) Role of predator eye size in risk perception by basking black iguanas (*Ctenosaura similis*). *Animal Behavior* 42:471–476

Burger J, Gochfield M, Murray BG (1992) Risk discrimination of eye contract and directness of approach in black iguanas (*Ctenosaura similis*). *Journal of Comparative Psychology* 106:97–101

Coricelli G, Nagel R (2009) Neural correlates of depth of strategic reasoning in medial prefrontal cortex. *Proceedings of the National Academy of Sciences USA* 106:9163–9168

Darwin C (1882) *The variation of animals and plants under domestication (2nd edition).* London: John Murray

De Waal FBM (1982) *Chimpanzee politics: Power and sex among apes.* Baltimore: Johns Hopkins University Press

Diamond J (2005) *Guns, germs and steel. A short history of everybody for the last 13,000 years.* London: Vintage

Dugatkin L, Trut L (2017) *How to tame a fox (and build a dog).* Chicago: The University of Chicago Press

Dunbar RIM, Shultz S (2007) Evolution in the social brain. *Science* 317:1344–1347

Dunbar R (2020) *Evolution. What everyone needs to know.* Oxford: Oxford University Press

Emery NJ (2000) The eyes have it: The neuroethology, function and evolution of social gaze. *Neuroscience and Biobehavioral Reviews* 24:581–604

Guo K, Robertson RG, Mahmoodi S, Tadmor Y, Young MP (2003) How do monkeys view faces?—A study of eye movements. *Experimental Brain Research* 150:363–374

Hare B, Brown M, Williamson C, Tomasello M (2002) The domestication of social cognition in dogs. *Science* 298:1634–1636

Hare B, Call J, Tomasello M (2001) Do chimpanzees know what conspecifics know? *Animal Behaviour* 61:139–151

Hare B, Plyusnina I, Ignacio N, Schepina O, Stepika A (2005) Social cognitive evolution in captive foxes is a correlated by-product of experimental domestication. *Current Biology* 15:226–230

Hare B, Wobber V, Wrangham R (2009) The self-domestication hypothesis: Evolution of bonobo psychology is due to selection against aggression. *Animal Behaviour* 83:573–585

Hecht EE, Smaers JB, Dunn WD, Kent M, Preuss TM, Gutman DA (2019) Significant neuroanatomical variation among domestic dogs breeds. *Journal of Neuroscience* 39:7748–7758

Kolling N, Braunsdorf M, Vijayakumar S, Bekkering H, Toni I, Mars RB (2021) Constructing others' beliefs from one's own using medial frontal cortex. *Journal of Neuroscience* 41:9571–9580

Kruska DCT (2007) The effects of domestication on brain size. In: Kaas JH (Ed.) *Evolution of Nervous Systems*, vol 3, pp. 143–153

Maguire EA, Gadian DG, Johnsrude IS, Good CD, Ashburner J, Frackowiak RSJ, Frith CD (2000) Navigation-related structural change in the hippocampi of taxi drivers. *Proceedings of the National Academy of Sciences USA* 97:4398–4403

Mars RB, Sallet J, Neubert FX, Rushworth MFS (2013) Connectivity profiles reveal the relationship between brain areas for social cognition in human and monkey temporoparietal cortex. *Proceedings of the National Academy of Sciences USA* 110:10806–10811

Perrett DI, Hiertanen JK, Oram MW, Benson PJ (1992) Organization and functions of cell responsive to faces in the temporal cortex. *Philosophical Transactions of the Royal Society B* 335:23–30

Rilling JK, Scholz J, Preuss TM, Glasser MF, Errangi BK, Behrens TEJ (2012) Differences between chimpanzees and bonobos in neural systems supporting social cognition. *Social, Cognitive and Affective Neuroscience* 7:369–379

Sallet J, Mars RB, Noonan MP, Andersson JL, O'Reilly JX, Jbabdi S, Croxson PL, Jenkinson M, Miller KL, Rushworth MFS (2011) Social network size affects neural circuits in macaques. *Science* 334:697–700

Saxe R (2006) Uniquely human social cognition. *Current Opinion in Neurobiology* 16:235–239

Scholz J, Klein MC, Behrens TEJ, Johansen-Berg H (2009) Training induces changes in white-matter architecture. *Nature Neuroscience* 12:1370–1371

Stolk A, Hunnius S, Bekkering H, Toni I (2013) Early social experience predicts referential communicative adjustments in five-year-old children. *PLoS ONE* 8L:e72667

Testard C, Larson MS, Watowich MM, Kaplinsky CH, Bernau A, Faulder M, Harshall HH, Lehmann J, Ruiz-Labides A, Higham JP, Montague MJ, Snyder-Mackler N, Platt ML, Brent LJN (2021) Rhesus macaques build new social connections after a natural disaster. *Current Biology* 31:2299–2309

Udell MAR (2015) When dogs look back: Inhibition of independent problem-solving behaviour in domestic dogs (*Canis lupus familiaris*) compared with wolves (*Canis lupus*). *Biology Letters* 11:20150489

Vijayakumar S, Hartstra E, Mars RB, Bekkering H (2021) Neural mechanisms of predicting individual preferences based on group membership. *Social Cognitive and Affective Neuroscience* 16:1006–1017

Wilkins AS, Wrangham RW, Fitch WT (2014) The "domestication syndrome" in mammals: A unified explanation based on neural crest cell behavior and genetics. *Genetics* 197:795–808

Zahn R, Moll J, Krueger F, Huey RD, Garrido G, Grafman J (2007) Social concepts are represented in the anterior superior temporal cortex. *Proceedings of the National Academy of Sciences USA* 104:6430–6435

Chapter 6. The chimpanzee and the human

Maslin (2016) has written extensively about the influence of climate change on human evolution. Human dispersal out of Africa is described by Stringer (2012). Changes in body and diet of early proto-humans are described by Lieberman (2014), changes in their social organization and associated cognitive processes by Dunbar (2016) and Tomasello (2014; 2016).

A Yale Tale (2005) *Fossil footprints.* Yale University: Peabody Museum of Natural History. Via Internet Archive Wayback Machine. https://peabody.yale.edu /exhibits/fossil-fragments/history/fossil-footprints

Boorman ED, Behrens TE, Rushworth MF (2011) Counterfactual choice and learning in a neural network centered on human lateral frontopolar cortex. *PLoS Biology* 9:e1001093

Braunsdorf M, Blazquez Freches G, Roumazeilles L, Eichert N, Schurz M, Uithol S, Bryant KL, Mars RB (2021) Does the temporal cortex make us human? A review of structural and functional diversity of the primate temporal lobe. *Neuroscience and Biobehavioral Reviews* 131:400–410

Catani M, Mesulam M (2008) The arcuate fasciculus and the disconnection theme in language and aphasia: History and current state. *Cortex* 44:953–961

Dunbar R (2016) *Human evolution: Our brains and behavior.* New York: Oxford University Press

Eichert N, Verhagen L, Folloni D, Jbabdi S, Khrapitchev AA, Sibson N, Mantini D, Sallet J, Mars RB (2019) What is special about the human arcuate fasciculus? Lateralization, projections, and expansion. *Cortex* 118:107–115

Hartogsveld B, Bramson B, Vijayakumar S, Van Campen AD, Marques JP, Roelofs K, Toni I, Bekkering H, Mars RB (2017) Lateral frontal pole and relational processing: Activation patterns and connectivity profile. *Behavioural Brain Research* 355:2–11

Herculano-Houzel S (2012) The remarkable, yet not extraordinary, human brain as a scaled-up primate brain and its associated cost. *Proceedings of the National Academy of Sciences USA* 109:10661–10668

Koechlin E, Ody C, Kouneiher F (2003) The architecture of cognitive control in the human prefrontal cortex. *Science* 302:5648

Lieberman D (2014) *The story of the human body: Evolution, health and disease.* London: Penguin

Mars RB, Sallet J, Neubert FX, Rushworth MFS (2013) Connectivity profiles reveal the relationship between brain areas for social cognition in human and monkey temporoparietal cortex. *Proceedings of the National Academy of Sciences USA* 110:10806–10811

Mars RB, Sotiropoulos SN, Pasingham RE, Sallet J, Verhagen L, Khrapitchev AA, Sibson N, Jbabdi S (2018) Whole brain comparative anatomy using connectivity blueprints. *eLife* 7:e35237

Maslin M (2016) *The cradle of humanity: How the changing landscape of Africa made us so smart.* Oxford: Oxford University Press

Murray EA, Wise SP, Graham KS (2017) *The evolution of memory systems: Ancestors, anatomy, and adaptations.* Oxford: Oxford University Press

Neubert FX, Mars RB, Thomas AG, Sallet J, Rushworth MFS (2014) Comparison of human ventral frontal cortex areas for cognitive control and language with areas in monkey frontal cortex. *Neuron* 81:700–713

Passingham RE, Stephan KE, Kötter R (2002) The anatomical basis of functional localization in the cortex. *Nature Reviews Neuroscience* 3:606–616

Passingham RE, Wise SP (2008) *The neurobiology of the prefrontal cortex: Anatomy, evolution, and the origin of insight.* Oxford: Oxford University Press

Rilling JK, Glasser MF, Preuss TM, Ma X, Zhao T, Hu X, Behrens TEJ (2008) The evolution of the arcuate fasciculus revealed with comparative DTI. *Nature Neuroscience* 11:426–428

Sierpowska J, Bryant KL, Janssen N, Blazquez Freches G, Romkens M, Mangnus M, Mars RB, Piai V (2022) Comparing human and chimpanzee temporal lobe neuroanatomy reveals modifications to human language hubs beyond the fronto-temporal arcuate fascicle. *Proceedings of the National Academy of Sciences USA* 119:e2118295119

Stout D, Toth N, Schick N, Chaminade T (2008) Neural correlates of Early Stone Age toolmaking: Technology, language and cognition in human evolution. *Philosophical Transaction of the Royal Society B* 363:1939–1949

Stringer C (2012) *The origin of our species.* London: Penguin.

Tomasello M (2014) *A natural history of human thinking.* Cambridge: Harvard University Press

Tomasello M (2016) *A natural history of human morality.* Cambridge: Harvard University Press

Van Essen DC, Dierker DL (2007) Surface-based and probabilistic atlases of primate cerebral cortex. *Neuron* 56:209–225

Vendetti MS, Bunge SA (2014) Evolutionary and developmental changes in the lateral frontoparietal network: A little goes a long way for higher-level cognition. *Neuron* 84:906–917

Zachos J, Pagani M, Sloan L, Thomas E, Billups K (2001) Trends, rhythms, aberrations in global climate 65 Ma to present. *Science* 292:686–593

Chapter 7. The bird, the mouse, and the human

The theory of mitochondrial origins is discussed in detail by Nick Lane (2005). Originally posed by Gould (1989), the idea of rerunning evolution is worked out in depth by Kershenbaum (2021). The evolution of birds from dinosaurs is described in Steve Brusatte's (2018) excellent overview. H.M.'s life is described by Suzanne Corkin (2014). Stanislas Dehaene's theory on reading in the brain is discussed in detail in his book Reading in the Brain *(2009).*

Aplin LM (2019) Culture and cultural evolution in birds: A review of the evidence. *Animal Behaviour* 147:179–187

Behrens TEJ, Muller TH, Whittington JCR, Shirley M, Baram AB, Stachenfeld KL, Kurth-Nelson Z (2018) What is a cognitive map? Organizing knowledge for flexible behavior. *Neuron* 100:490–509

Brusatte S (2018) *The rise and fall of dinosaurs.* New York: William Marrow.

Changizi M (2010) *The vision resolution. How the latest research overturns everything we thought we knew about human vision.* Dallas: BenBella Books.

Clayton NS, Dickinson A (1998) Episodic-like memory during cache recovery by scrub jays. *Nature* 395:272–274

Clayton NS, Dally JM, Emery NJ (2007) Social cognition by food-caching corvids. The western scrub-jay as a natural psychologist. *Philosophical Transactions of the Royal Society London B* 362:507–522

Constantinescu A, O'Reilly JX, Behrens TEJ (2016) Organizing conceptual knowledge in humans with a gridlike code. *Science* 352:1464–1468

Corkin S (2014) *Permanent present tense: The man with no memory, and what he taught the world.* London: Penguin

Dehaene S, Cohen L (2007) Cultural recycling of cortical maps. *Neuron* 56:384–398

Dehaene S (2009) *Reading in the brain: The new science of how we read.* New York: Penguin.

Dehaene S, Pegado F, Braga LW, Ventura P, Nunes Filho G, Jobert A, Dahaene-Lambertz G, Kolinsky R, Morais J, Cohen L (2009) How learning to read changes the cortical networks for vision and language. *Science* 330:1359–1364

Gould SJ (1989) *Wonderful life. The Burgess Shale and the nature of history.* New York: W.W. Norton

Kershenbaum A (2021) *A zoologist's guide to the galaxy. What animals on earth reveal about aliens—and ourselves.* London: Penguin.

Lane N (2005) *Power, sex, suicide. Mitochondria and the meaning of life.* Oxford: Oxford University Press

Luongo FJ, Liu L, Ho CLA, Hesse JK, Wekselblatt JB, Lanfranchi FF, Huber D, Tsao DY (2023) Mice and primates use distinct strategies for visual segmentation. *eLife* 12:e74394

Moser MB, Rowland DC, Moser EI (2015) Place cells, grid cells, and memory. *Cold Spring Harbor Perspective in Biology* 7:a021808

Murray EA, Wise SP, Graham KS (2017) *The evolution of memory systems: Ancestors, anatomy, and adaptations.* Oxford: Oxford University Press

O'Keefe J, Nadel L (1978) *The hippocampus as a cognitive map.* Oxford: Oxford University Press

Olkowicz S, Kocourek M, Lucan RK, Portes P, Fitch WT, Herculano-Houzel S, Nemec P (2016) Bird have primate-like numbers of neurons in the forebrain. *Proceedings of the National Academy of Sciences USA* 113:7255–7260

Park SA, Miller DS, Boorman ED (2021) Inferences on a multidimensional social hierarchy using a grid-like code. *Nature Neuroscience* 24:1292–1301

Raby CR, Alexis DM, Dickinson A, Clayton NS (2007) Planning for the future by western scrub-jays. *Nature* 445:919-921

Shimizu T, Hinozuka K, Uysal AH, Kellogg SL (2017) The origins of the bird brain: Multiple pulses of cerebral expansion in evolution. In: Watanabe S et al. (Eds.), *Evolution of the brain, cognition, and emotion in vertebrates.* Brain Science.

Sridhar H (2017) Revisiting Clayton and Dickinson 1998. *Reflections on papers past: Revisiting old papers in ecology and evolution through interviews with their authors* https://reflectionsonpaperspast.com

Stacho M, Herold C, Rook N, Wagner H, Axer M, Amunts K, Güntürkün O (2020) A cortex-like canonical circuit in the avian forebrain. *Science* 369:eabc5534

The Avian Brain Nomenclature Consortium (2005) Avian brains and a new understanding of vertebrate brain evolution. *Nature Reviews Neuroscience* 6:151–159

Thiebaut de Schotten M, Cohen L, Amemiya E, Braga LW, Dehaene S (2014) Learning to read improves the structure of the arcuate fasciculus. *Cerebral Cortex* 24:989–995

IMAGE CREDITS

0.1 Mammalian brains based on originals from the University of Wisconsin and Michigan State Comparative Mammalian Brain Collections, as well as from those at the National Museum of Health and Medicine (brainmuseum.org). Preparation of all the original brain images and specimens from brainmuseum.org was funded by the National Science Foundation, as well as by the National Institutes of Health. Corvid brain based on figure 1A of Olkowicz et al. (2016) Birds have primate-like numbers of neurons in the forebrain. *Proceedings of the National Academy of Sciences* USA 113:7255–7260. Lamprey and dragon brains based on figure 1 of Northcutt RG (2002) Understanding vertebrate brain evolution. *Integrative and Comparative Biology* 42:743–756.

1.0 Based on figure 1 of Northcutt RG (2002) Understanding vertebrate brain evolution. *Integrative and Comparative Biology* 42:743–756.

1.2 Lamprey brain based on figure 1 of Northcutt RG (2002) Understanding vertebrate brain evolution. *Integrative and Comparative Biology* 42:743–756.

2.0 Left side based on figure 1 of Northcutt RG (2002) Understanding vertebrate brain evolution. *Integrative and Comparative Biology* 42:743–756. Right side based on the University of Wisconsin and Michigan State Comparative Mammalian Brain Collections, as well as from those at the National Museum of Health and Medicine (brainmuseum.org). Preparation of all the original brain images and specimens from brainmuseum.org was funded by the National Science Foundation, as well as by the National Institutes of Health.

2.1 Left side based on Murray EA, Wise SP, Graham KS (2017) *The evolution of memory systems.* Oxford: Oxford University Press. Right side based on Molnar Z (2011) Evolution of cerebral cortical development. *Brain, Behavior and Evolution* 78:94–107

223

3.0, 4.0, 5.0, 6.0 Based on originals from the University of Wisconsin and Michigan State Comparative Mammalian Brain Collections, as well as from those at the National Museum of Health and Medicine (brainmuseum. org). Preparation of all the original brain images and specimens from brainmuseum.org was funded by the National Science Foundation, as well as by the National Institutes of Health.

3.2 Penfield map from "The 'motor homunculus.' The body surface is projected onto the gyrus precentralis (coronal section of the right hemisphere)," created by ralf@ark.in-berlin.de and downloaded from Wikimedia Commons under CC BY-SA 4.0 license (https://creativecommons.org /licenses/by-sa/4.0/deed.en)

3.3 Reproduced from "Raven's Progressive Matrices Example" by Life of Riley and downloaded from Wikimedia Commons under CC BY-SA 3.0 license (https://creativecommons.org/licenses/by-sa/3.0/deed.en)

4.2 Reproduced from figure 1 of Van der Linden M, Murre JMJ, van Turennout M (2008) Birds of a feather flock together: Experience-driven formation of visual object categories in human ventral temporal cortex. *PLoS ONE* 3:e3995, published under the terms of the Creative Commons Attribution License.

6.1 Modified from figure 2b of Rilling JK, Glasser MF, Preuss TM, Ma Z, Zhao T, Hu X, Behrens TEJ (2008) The evolution of the arcuate fasciculus revealed with comparative DTI. *Nature Neuroscience* 11:426–428

6.2 Reproduced from figure 4a of Mars RB, Sotiropoulos SN, Passingham RE, Sallet J, Verhagen L, Khrapitchev AA, Sibson N, Jbabdi S (2018) Whole brain comparative anatomy using connectivity blueprints. *eLife* 7:e35327.

7.0 Rat brain based on original from the University of Wisconsin and Michigan State Comparative Mammalian Brain Collections, as well as from those at the National Museum of Health and Medicine (brainmuseum.org). Preparation of all the original brain images and specimens from brainmuseum.org was funded by the National Science Foundation, as well as by the National Institutes of Health. Corvid brain based on figure 1A of Olkowicz et al. (2016) Birds have primate-like numbers of neurons in the forebrain. *Proceedings of the National Academy of Sciences* USA 113:7255–7260.

7.1 Modified from figure 1 of Moser MB, Rowland DC, Moser EI (2015) Place cells, grid cells, and memory. *Cold Spring Harbor Perspectives on Biology* 7:a021808

INDEX

Note: Page numbers followed by an *f* refer to figures.

dodo, 4–5
dogs: brain structure of, 119;
    domestication of, 125; problem
    solving in, 130; research on, 85–86,
    128, 131–132
domestication and domesticated
    animals: process of, 123–132;
    self-domestication, 133
dopamine neurons, 35–36, 37f, 38, 149
dorsal cortex: function of, 55–56;
    structure of, 54f, 56
dorsal pallium, 190
Dubois, Eugene, 5–7, 41, 150, 156, 186
Dunbar, Robin, 136–137
Duncan, John, 92–93
dyadic relationships, 136

earth, history of life on, 18–22
East African Rift Valley, 151–153
Eastern grey squirrel, 69, 74–75
ecology, 10
Ediacaran period, 20
Edinger, Ludwig, 190
effectors, 86–87
eggs, 47
electrical imbalance, 23
elephants, 68
emus, 189
encephalization, 7
endocasts, 7
end-Permian extinction, 48
entorhinal-hippocampal system,
    197–200
environment, information about,
    29–30
Eocene period, 100, 102
episodic memory, 112, 184–185
ethics, 164, 166
eukaryotic cells, 179
evolution: of brains, 181–192;
    convergent evolution, 9;

domestication and, 125;
    evolutionary tree, 8f, 9, 179,
    205–206; history of, 18–22; laws
    of, 180; process of, 206; theory
    of, 3–5, 186
expertise, 96–97
extinctions, 44–52, 187
eye movements, 88
eyes and eye gaze, 71, 119–122, 140,
    176

face processing, 107–108, 111, 139
false belief paradigm, 140–141
feathers, 187
feature detection cells, 105–106
feeding habits. See foraging; hunting
fins, 45, 52–53
fish, 22, 45
flight, 188, 191–192
food and food webs, 20, 101
foraging: adaptation and, 101–104;
    brain cells and, 30–38; coopera-
    tion and, 122; food caches and,
    184; frontal cortex and, 113–118,
    173–174; fruit, 102; humans and,
    157–162; hunter-gatherer strategy,
    159; natural selection and, 17–18;
    predictive foraging, 33–34;
    process of, 31; strategies, 64; use
    of tools, 158
forebrain: function of, 29; structure
    of, 27f, 190
fornix, 182
fossils and fossil record: Darwin's
    observations on, 3; evolution and,
    20; feathers, 187; humans and, 151;
    primates and, 72, 98; study of, 6–7;
    teeth, 158; therapods, 188;
    Tiktaalik, 45–46, 186
fovea, 102
foxes, 123–124, 126–128, 130